COMPUTERS IN QUANTITY SURVEYING

COMPUTERS IN QUANTITY SURVEYING

R. J. ALVEY F.R.I.C.S., F.I.Q.S.

Principal Lecturer in Quantity Surveying,
Trent Polytechnic

First published 1976 by
THE MACMILLAN PRESS LTD
London and Basingstoke
Associated companies in New York Dublin
Melbourne Johannesburg and Madras

Paper edition ISBN 333 17973 0
Cased edition ISBN 333 21397 1

Set in IBM Press Roman by
PREFACE LTD
Salisbury, Wilts
Printed in Great Britain by
UNWIN BROTHERS LTD
Woking

This book is dedicated to my wife Beryl,
for her patience during its preparation

CONTENTS

PREFACE

Quantity surveyors are becoming increasingly involved in modern methods of data processing and are looking to the computer as a means of solving problems and providing techniques for the more effective use of their skills. The computer provides new techniques that the construction design team can utilise to the betterment of the design and construction processes. The changing economic climate and the introduction of new building techniques have demanded a closer liaison between the professions and the construction team with the object of providing coordinated information systems. Computerisation may be held to be an answer. A study of computers, therefore, is necessary if the quantity surveyor is to keep abreast of modern techniques.

Computer jargon can be very confusing and only confounds the layman; this tends to create an air of mystery. Nevertheless the introduction of computers has created a new professional with an expertise that the quantity surveyor cannot hope to replace. However, to utilise these new skills the quantity surveyor must have some knowledge of the techniques involved. This means that he should have a basic appreciation of the technicalities of computerisation and a knowledge of its potential applications.

This book is directed primarily at students who are preparing for the professional examinations of the Royal Institution of Chartered Surveyors, the Institute of Quantity Surveyors and the Institute of Building. It is hoped that it may also be found useful by students studying for degrees and diplomas in quantity surveying and building and for allied professions and practitioners.

The book makes a study of computers to provide a basic appreciation of their nature and characteristics and to give a better understanding of the manner in which computers may be used satisfactorily by the construction design team with particular reference to the quantity surveyor's needs.

I wish to acknowledge with gratitude all the help and cooperation given by many organisations and persons associated with the computer industry especially R. Aldridge, D. R. Judd and S. K. A. Raza and many others too numerous to mention. My special thanks are due to my colleagues at Trent Polytechnic for their encouragement, especially to Dr I. H. Seeley for his most helpful comments. Grateful thanks are also due to those quantity surveyors with whom I have had some stimulating discussions. Last and by no means least I wish to thank my family for their tolerance and understanding while this book was being prepared.

Nottingham,   R. J. ALVEY
Autumn 1975

1 THE IMPACT OF THE COMPUTER

To err is human.
To really mess things up requires a computer.
                              (Bill Vaughan)

Computers are having an ever-increasing influence on our lives and it is almost impossible to ignore them. Our wages, salaries or fees may have been processed by computer. Our bank statements and demands for payment will probably also have been generated by a computer. Computers seem to leave their mark everywhere from space projects to the latest rates demand. In fact we appear to have reached a stage when everything seems to be computerised or is going on the computer and its influence seems inescapable. To what extent, if at all, does the computer mess things up? Stories abound regarding the computer's apparent bungling: the domestic gas bills for excessive amounts, or the discharge of an airman from the Air Force because he was pregnant. Some are possibly true, others pure fantasy. Derogatory stories do circulate about things that are clouded in mystique and not clearly understood. It is hoped that the following chapters therefore will help lift the veil from what may be a clouded picture of computers and put it into perspective.

To mention computers seems to conjure up in the mind certain pictures of giant automatons that will eventually control man and his environment. Our minds boggle at man's ingenuity and we may become confused by the computer's intellectual attainments. To define the attributes of a computer is difficult. The computer's ideal properties may be listed and we may say that it is an extension of man's thinking: a so-called giant brain that will make the human brain obsolete. We could consider the argument as to whether computers can think and have other attributes such as a capacity for learning. On the other hand the computer may be described as a tool or an instrument for man's use, which will carry out his instructions to the letter.

THE COMPUTER'S ABILITIES

Computers seem to display many of the characteristics associated with human behaviour. They perform logical operations like comparing and choosing alternatives, matching equals or selecting the next instruction to be carried out. They also have the ability to remember and to make what appear to be logical decisions. In order to appreciate the complex behaviour of computers we must remember that they lack the power of critical judgement and the capacity for creative thinking. The computer has a built-in set of rules that is steadfast, and it performs operations on given information without questioning the factual truth of the data to be processed or the validity of its operations. It will blindly

1

carry out operations in accordance with the instructions it has been given and as such the computer may be described as an 'electronic idiot' rather than an 'electronic brain'. Developments in science and mathematics have occurred by questioning axioms and rules that were thought to be self-evident. No computer can do this, nor can it use past accumulated knowledge to arrive at new possibilities and inventions. Computers will remain robots so long as the highest form of human thinking cannot be duplicated.

The one important quality that puts the modern computer in a class by itself is its ability to operate automatically. A long sequence of related operations can be performed according to a predetermined program of instructions without the need for human intervention. What it will do then is operate automatically without questioning its instructions. The computer has been used to play games such as chess. To play a game of chess with a computer and win depends on the instructions that the chess 'master' has given. Failure of the computer to win may be a result of inadequate instructions necessary to counteract certain moves.

Information expansion is now made possible on a scale that was not envisaged before the introduction of electronic computers. Records can now be kept with the possibility of retrieval and updating in a very short space of time. Events may be recorded and reported as soon as they happen. The computer can lay claim to features that are lacking both in the human brain and the more conventional methods of data processing. Business activity is toned up since the computer does not tolerate slipshod or inaccurate data preparation and it is essential that 'good housekeeping' be maintained.

BENEFITS OF USING A COMPUTER

The general benefits that may be gained by using a computer may be summarised as follows.[1]

(a) *Accuracy*   The computer can handle data and continue to handle it in such a manner as to maintain accuracy. Human beings are prone to making mistakes as they become tired but the computer without human intervention will maintain its standard of performance. It should be appreciated that this is dependent upon the accuracy of the data that are input and which conform to the system requirements. Errors usually occur as a result of human fallibility rather than the operation of the machine.

(b) *Speed*   A special feature of the electronic computer is its speed of operation. Calculations are performed and decisions made in fractions of a second; a millionth of a second is not unusual. This is something that human beings are incapable of doing or even comprehending.

(c) *Retrieval of Information*   The computer can maintain full historical records with a rapid access. Masses of information can be stored in a very small space. Considerable research is being carried out with a great measure of success into the construction of memory units where information is stored by using laser beams both to read and write information.

(*d*) *Handling Complex Problems*   Tasks may be performed that are of almost infinite complexity, in fact as complex as the human brain can comprehend. If it were not for computers, man's progress into outer space would be almost impossible.

These benefits are what make the computer unique in the business world and make it possible to control certain areas of business activity that have been previously uncontrolled because of the excessive cost of clerical labour.

IMPACT OF THE COMPUTER

From the benefits to be gained by using an electronic computer it can be seen that its usefulness extends into many areas such as research, design and production, commerce and administration. The breakthrough in computation represents a great achievement in devising better methods of performing repetitive and routine tasks more efficiently. Many people feared that the introduction of computers would bring mass unemployment. However, experience has shown that the computer's impact has created an entirely new job market in computer technology and operation.

The areas of application range from record-keeping, management information systems, information storage and retrieval, electronic guidance and control, and model simulation to computer-aided design.[2]

Record-keeping   Although computers were not developed for this particular purpose they are committed to processing record-keeping applications such as payroll and inventory accounting, production scheduling and invoicing. Banks see electronic data-processing as a means of increasing efficiency by permitting continual updating of depositors' accounts. Industrial enterprises and government departments require records of accounts payable and receivable. Airports also use computers to maintain records of bookings.

Management Information Systems   The computer can be used to provide an all-inclusive system designed to meet the instant information needs of a management as a means of increasing the efforts and efficient operations of a business. Data coordination is an important factor receiving much attention in the construction industry and the computer is seen as a means of achieving this end.[3]

Information Storage and Retrieval   This is an important feature of some management operations. The computer can be used to assist in medicine: for instance the observation of patients is made possible by means of electronic equipment. Doctors can now use the computer to test and diagnose various human ailments and the National Health Service has gained advantages by using the computer for storage and retrieval of patients' records. However, the use of the computer in this field is still in its infancy.

Education has also gained by adopting computerisation. Electronic computers contribute to new and revolutionary ideas such as teaching aids. Computers were introduced into education mainly for computerised programmed instruction.

3

However, in recent years the computer structure, its technology and computer languages have been taught as a computer science.

The police force is a well-known area where computerised data storage and retrieval are found to be very useful. Data are built up and stored in the computer in such a way as to provide a readily accessible source of information. VASCAR, the Visual Aid Speed Computer and Recorder, is a new computerised device whereby the police may detect the speed of traffic.

Libraries have problems of disseminating knowledge on a wide scale and collecting information from many and varied sources. The answer is provided by the computer as a tool for implementing a communications network. The Lloyds Shipping Register is an example of a type of information retrieval system with information held by the computer and retrieved for printing quick answers to queries. Updating is made easier if information is filed in this manner.

Electronic Guidance and Control   The industrial use of computers is generally associated with automation. The manufacturing processes have utilised computers for the control of machine tools. Process control by the computer may also be found in the chemical industry for the production of such commodities as ethylene, nylon, rubber and ammonia.

A better-known area where computers are used for guidance and control is aeronautics. Manned space flights are made possible by computers that have performed the initial calculations, which may take hundreds of man-hours to solve. Split-second calculations are also necessary to maintain flight. These can only be performed at the required speed by the computer.

Model Simulation   Simulation is the experimental technique of operating and studying the model of a system by an empirical method. This means that in order to try out the multitude of variables a mathematical or physical model must be constructed. Simulation therefore refers to a symbolic representation for learning about and testing an idea before making it operational. Computer techniques make simulation an easier process. The simulation of flight by computer has been used for both designing aircraft and the training of pilots. Business executives are able to simulate conditions requiring certain key decisions relating to a business activity. Models may also be developed using advanced mathematical techniques and linear programming to simulate traffic movements and the distribution of work.

THE USE OF THE COMPUTER IN THE CONSTRUCTION INDUSTRY

Like many other industries the construction industry has not escaped the impact of the computer. Architects, consulting engineers, contractors, subcontractors and quantity surveyors are utilising the computer's facilities.

The Computer and the Architect

Architects are using the facilities provided by modern computing techniques to provide basic design information. Architectural and engineering design may involve the performance of tedious calculations for a wide range of variables to permit a greater measure of choice among feasible designs. A number of environmental factors relating to building design, such as heat losses and lighting, may conveniently be processed by computer. Use is also made of the processing facilities for scheduling and costing.

It is now possible to use computers in the design process. The development of computer technology now permits the architect and engineer to view completed designs in three dimensions on a cathode-ray tube. A number of programs are available by which the architect can produce drawings using computer graphic displays.[4,5,6] Computer graphic displays involve the use of visual display units using cathode-ray tubes on to which drawings may be projected. The designs may be created in three dimensions and rotated for perspective views. Drawings may then be produced in a number of different displays under computer control.[4,5] Systems to extend the graphic design are being developed so that a costing and the generation of drawings, schedules and quantities may be produced from the basic raw data.

The use of the computer by the architect means that more designs can be examined and assessed with the added advantage of reaching an optimum solution. In this way a more rational approach can be taken towards the layout of a building. Programs performing this type of work may be considered as a bonus for those who look upon the use of a computer for harder economic reasons. Computer output makes for easier communications between designer and client and the rest of the design team with an increased speed of operation in design resulting in an earlier building completion.

The Computer and the Contractor

With the increased competition in the construction industry there is a need for more sophisticated methods of processing and the contractor is turning to the computer as a tool to ensure a profitable performance.

Computer systems are being used in many areas of the contractor's project control such as job-costing, project-planning, scheduling and cost control.[7] The contractor may find the computer of use in assisting him with such problems as payroll, purchasing, invoicing and stock control, information retrieval and many scheduling operations.[8] Computers provide an opportunity to simulate a project and compare alternative proposals to arrive at a satisfactory solution. Particular value is gained by this exercise for assessing the alternatives available when project-scheduling. Project cost control can be implemented using computer programs that will produce cost information relating the actual costs to estimated costs. The contractor needs progress information relating to a project at regular intervals. This can be supplied satisfactorily by the computer with a saving of manual effort.

One of the more tedious accounting operations that the contractor has to perform is the handling of subcontractors' accounts. These need to be broken down into headings such as measured work, materials on site, daywork, fluctuations and other claims, each of which may have an applicable discount and retention. Every month, or sometimes more often, the appropriate discounts, retention and any contra-accounts must be applied to each relevant subcontractor's accounts and the necessary calculations made before payment.

The computer can be used to apply validity checks and perform all the necessary calculations to give the amount of payment due on certified valuations together with a print-out.

Contractors who operate plant departments that carry many spare parts or require bulk stock holdings can benefit by applying a computer program to their inventory control. This provides a method of controlling the replacement of stocks in such a way as to minimise the costs of holding too much stock and obsolescent parts. The whole of the stores issue and accounting procedure can be operated in conjunction with inventory control. The industrial sector of the construction industry makes use of a computer chiefly as an aid to management. Manufacturers, however, use it as a means of process control. These applications can be of use in market research, sales analysis and sales forecasting.

The Computer and the Quantity Surveyor

The production of bills of quantities forms one of the prime functions of the quantity surveyor and is a starting point for his other analytical activities. It is from the raw data that information can be extracted for other uses. A working party was set up by the Quantity Surveyors' Committee of the Royal Institution of Chartered Surveyors in February 1960 to look into the possibilities of the use of computers in connection with working up quantity surveyors' dimensions to streamline and quicken up the process.[9],[10] It recommended an investigation into data processing for the production of bills of quantities. The Committee also suggested that the use of the computer for such purposes might act as a catalyst for speeding up improvements in planning and constructional methods of the construction industry.

The use of the computer brings a logical discipline to quantity surveying procedures. Also many benefits may accrue as a 'spin-off' from computer processing as an extra to the normal requirements. The quantity surveyors' principal use of the computer is for storing and resorting data, scanning large files of information and retrieving what is relevant. He is able to provide additional information to serve a number of aspects of a building project. For instance bills of quantities may be produced in various formats from the same source data. The computer may be used to provide contractors' information for use in all features of contract and site management including cost planning and design optimisation from the quantity surveyors' measured work.

The quantity surveyors' work processed by the computer varies from little more than calculating and collating coded descriptions and dimensions to more elaborate systems involving processes for the retrieval of information that would otherwise be buried in the quantity surveyors' take-off. The following chapters examine some of the systems and techniques used to achieve these ends.

REFERENCES

1. T. F. Fry, *Computer Appreciation* (Butterworth, London, 1970).
2. E. M. Awad, *Automatic Data Processing* (Prentice-Hall, Englewood Cliffs, N.J., 1970).
3. —— General Brochure, Construction Control Systems Ltd.

4. B. Auger, *The Architect and the Computer* (Pall Mall, London, 1972).
5. T. W. Maver, 'PACE 1 Computer Aided Building Design', *Architects' J.* 28 July (1971).
6. —— *Computer Facilities and Techniques*, Integrated Building Information Development.
7. J. B. Bonny and J. P. Frein, *Handbook of Construction Management and Organization* (Van Nostrand Reinhold, New York, 1973).
8. W. R. Martin, *Network Planning for Building Construction* (Heinemann, London, 1969).
9. Royal Institution of Chartered Surveyors, Quantity Surveyors' Committee, 'The Use of Computers', *Chart. Surv.*, 93 (1961). pp. 561–3
10. Royal Institution of Chartered Surveyors, Quantity Surveyors' Committee, 'Use of Computers for Working Up', *Chart. Surv.*, 94 (1961) pp. 248–58.

2 THE NATURE OF THE COMPUTER

Computerisation, for all its apparent complexity, is surprisingly straightforward when examined in detail. Like many other products of human ingenuity, the computer is composed of a number of individual units each of which has a task to perform and it is in the terms of these tasks that the computer can be described. However, before computers are examined in any detail it is important to understand their nature and characteristics. The basic types of computer in use today may be described under the following headings.

(a) Analog computers
(b) Digital computers
(c) Hybrid computers

ANALOG COMPUTERS

The analog computer is a device that measures the quantity of a continuous variable physical condition. In other words a measure of 'how much'. For instance speed, temperature and time are physical conditions that vary in their amount and are continuously changing. The measure of 'how much' is represented by analogy in the form of some visual representation.

The speed of a car changes with acceleration and deceleration and is a variable physical condition that we can experience. The measure of how fast the car is travelling is indicated by the position of a needle on the dial of the speedometer. Heat is a physical condition that is experienced by everyone. This is purely comparative in the way it is experienced: if we plunge our hands into a bowl of cold water and then immediately into a bowl of water at body temperature, the second bowl of water will seem hot to our senses. A thermometer will indicate the amount of heat (temperature) by a visual representation on a scale marked out in degrees. *'Tempus fugit'* (time flies), but just how quickly does it pass? The measurement of time is by the analogy of the position of hands on a clock to represent a point in time; although devices have been developed that rely on a counting action and represent time in terms of numbers — these are called digital clocks and should not be confused with the conventional timepiece, which measures time by analogy.

The measuring devices that compute a measure of 'how much' by analogy are known as analog computers. A measure of 'how much' has its own application and each analog computer is designed to perform a specific task. The quantity surveyor is concerned with a measure of 'how much' when he computes from a scale of distance such as the tape measure or scale rule. Our earliest forefathers were constantly making analogies; for example, the distance between one outstretched arm and the other, or the distance from the tip of a man's nose to his outstretched middle finger were units of measurement that he used to

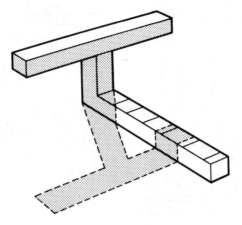

Figure 2.1   Analog device – early Egyptian crossbar

represent a measure of 'how much'. A more scientific unit of measurement used by surveyors today is the metre, which is one ten-millionth part of the distance between the equator and the North Pole. Engineers and surveyors have also been using the slide rule for some time as a means of computing quantities by the analogy of scales of distance.

An analog computer can then be defined in general as a calculating device that indicates by analogy the amount of some physical condition on a scale of measurement. Thus any measuring device that solves problems by translating a physical phenomenon into related quantities will come under this heading. The early Egyptians used a crossbar (figure 2.1) and the motion of the sun across the sky to measure daylight hours;[1] this was the forerunner of the sundial. At dawn the crossbar was set facing east, the shadow cast by the crossbar indicated the six hours before noon. At noon, when the shadow was the shortest, the crossbar was turned round. The declining hours were then measured with the crossbar facing west. Sundials are similar in some ways to this early device although when seen today they act more as a decoration than a timepiece.

DIGITAL COMPUTERS

The digital computer is concerned with a measure of 'how many'. The problem to be solved consists of the handling and processing of discrete items. The computation is performed on a numerical mathematical basis consisting of the adding, subtracting, dividing and multiplying processes. Our earliest forefathers had to rely on counting devices to enable them to assess 'how many'. In the very early days they used their fingers to count up to ten. More sophisticated methods involved the use of arms and bending fingers up in various ways to perform multiplication problems. Markings used as counting devices can be seen on walls of prehistoric caves. The early Egyptian merchants made a device using pebbles in grooves in the sand; a pebble in the right-hand groove represented a unit of one; when ten pebbles were counted in the right-hand groove they

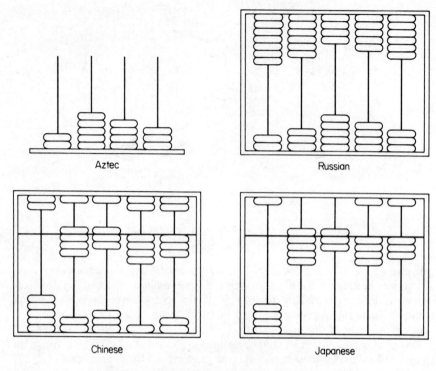

Figure 2.2   Digital device – abacus

exchanged them for one pebble placed in the groove on the immediate left. Each groove represented units, tens, hundreds or thousands, and so on.

A more sophisticated device is the *abacus*.[2] This was used in widely separate cultures and appears to have been invented independently in several centres as is evidenced by such versions as the Aztec, Russian, Chinese and Japanese forms, as indicated in figure 2.2. Some prefer to use the abacus and claim that it can be used as quickly as most modern counting machines. Its use, however, has been superseded by more modern devices that can be either mechanical or electronic. With the aid of digital computers the quantity surveyor is now relieved of the mental gymnastics of squaring and casting his dimensions.

The measure of 'how much' and 'how many' can best be illustrated by looking at the dashboard of a car. The speedometer is the analog device, which shows the speed of the car ('how much') and the milometer shows the number of miles travelled ('how many'); the milometer then is the digital device. The simple principle for distinguishing between analog and digital computing devices depends on whether the computation sets up an analogy of the problem or whether it is performed by the numerical counting of discrete items. All have characteristic states or conditions of activity that can be compared to illustrate some of their distinguishing features. Some of these features are compared as follows.[3]

(a) The main distinguishing feature is that the analog device sets up an analogy of the problem – a question of 'how much' – while the digital device breaks the problem down into arithmetic – a question of 'how many'.

(b) Measurements on an analog device are variable and continuous and are translated by means of the rotation of a shaft, the amount of voltage or the position of an indicator on a scale of measurement. Measurements on a digital device are discontinuous and separate from the problem and are represented by numbers in a discrete pattern.

(c) The analog device is designed for a specific purpose and the basic operation is performed by a single-purpose device such as the speedometer in a car. The devices used for digital purposes may be composed of interchangeable units working together. The operations are carried out by a combination of a number of devices such as adders, registers, accumulators, and the like.

(d) Analog devices serve as models and reflect the relationship of measurement with an actual physical quantity and carry out their operations in the actual time of the physical condition. Digital devices compound arithmetic data that are unrelated to the system they represent and the time of the operations do not correspond to real time.

(e) The analog device can only handle measurements of a continuous variable nature and is best suited to simulate a response to a physical condition by a mathematical analogy. Digital devices handle data and numeric problems of a business or scientific nature that involve discrete random processes.

HYBRID COMPUTERS

The type of computer that makes use of both the analog and the digital components and methods is knows as a hybrid computer.[2,3] This type of computer possesses the speed, flexibility and direct communication of the analog computer with the logic, memory and accuracy of the digital computer. The digital computer can be adapted to represent a continuous variable function and characteristics of an analog computer. For instance an arrangement of discrete characters can be used to create some graphical representation such as a bar chart (see chapter 10). A number of portraits have been produced using this method, and competitions have been held for the best portrait so produced. These results are not obtained by the use of hybrid computers but rather by the output of a digital computer, the pictorial effect being obtained through the arrangement of discrete characters. However, a digital computer can be made to produce analog results by equipping it with devices to convert input from an analog to a digital form and output from a digital to an analog form.

ELECTRONIC COMPUTERS

In recent years modern developments in technology and electronics have seen the advent of data-processing devices that are replacing the electromechanical calculating machines. It is this area that most people associate with computers. The term electronic computer bears reference to the technology adopted; the machine consists of electronic circuits and components through which pulses of electricity flow, which represent data to be processed.

11

The first automatic computer called an Automatic Sequence Controlled Calculator, which is alleged to have been built at Harvard University, was put into service in 1944. It put into practice many of the concepts formulated by Charles Babbage one hundred years or so earlier and made use of electromechanical techniques. The first all-electronic computer, ENIAC (Electronic Numerical Integrator and Calculator), was built at the University of Pennsylvania in order to solve problems of ballistics and aeronautics for the United States Army in 1946.[2] General-purpose electronic computers began to appear in the United States in 1948, and by 1960 their use was well established. UNIVAC 1 (Universal Automatic Calculator) was the first commercial digital computer produced in the early 1950s.[4,5]

Electronic technology relating to computers is classified by reference to a specific 'generation'. This is a means of classifying the computer's technical characteristics.[6]

First Generation   These particular types of machine were operating during the years 1954 to 1959. They worked by means of circuits consisting of wires and thermionic valves. Compared with present-day computers they were large in size and tended to fail frequently due to the heat that was generated. Processing was in the millisecond speed range (one-thousandth of a second) with a comparatively low internal storage. All models tended to have individual characteristics and lacked any relationship one with another.

Second Generation   The circuits of wires and the thermionic valves of the first-generation machines were replaced by printed circuits, diodes and transistors and such machines operated during the years 1959 to 1964. The advances in technology enabled smaller machines to be built with a subsequent reduction in the generation of heat. This increased the degree of reliability since transistors and solid-state components do not have such a high failure rate as thermionic valves. The internal storage capacity was also increased in size. The processing time was increased to a microsecond (one-millionth of a second) with the addition of direct access storage (see chapter 3).

Third Generation   In 1964 the technology of microminiaturisation was introduced and computers were built with micro-integrated circuits, which comprised integrated components rather than individual components with soldered connections. The introduction of these integrated circuits resulted in higher processing speeds of a nanosecond (one-thousandth of a microsecond). Even smaller units can be produced resulting in a higher capacity internal storage with additional facilities for multiprogramming and remote communication (see chapter 3).

MINICOMPUTERS

A further development in computer technology is the minicomputer and its associated peripherals. These have been made possible by the advances in space research and telecommunications coupled with the miniaturisation of components. The development and mass production of integrated circuits have made possible the manufacture of smaller machines at greatly reduced costs.

The minicomputer is a digital computer that uses short 16-bit words (see chapter 3) to represent data and computer instructions.[7] Minicomputers are portable and work especially well with conversational terminals and graphic displays. They can also serve as special-purpose computers in control, instrumentation and communications.

The advantages of using a minicomputer may be summarised as follows.[8]

(a) The minicomputer may be installed at a cost that is considerably lower than a main-frame computer of equivalent power.
(b) The minicomputer may be installed easily in any position where there is a power point.
(c) The operation of the minicomputer does not require organisational structures or specialist staff.
(d) The minicomputer can be used either as a terminal to a larger machine or locally as a processor in its own right.

The initial development of the minicomputer took place in the late 1950s out of a need for low-cost process control computers. It is the low cost of a minicomputer and its processing power that has created such an appeal. Osborne[5] asserts that 'the future of the computer industry belongs to the minicomputer, the prefix "mini" no longer applies to computing power but rather to small physical size'. The demand for minicomputers has created an inroad into the computer market and it is competing keenly with its larger counterpart, the main-frame computer.

REFERENCES

1. A. Vorwald and F. Clark, *Computers from Sand Table to Electronic Brain* (Lutterworth, Guildford, 1966).
2. S. H. Hollingdale and G. C. Tootill, *Electronic Computers* (Penguin, Harmondsworth, 1966).
3. H. Jacobowitz, *Electronic Computers Made Simple* (W. H. Allen, London, 1967).
4. E. M. Awad, *Automatic Data Processing* (Prentice-Hall, Englewood Cliffs, N. J., 1970).
5. —— *The Value of Power* (General Automation Inc., Osborne, London, 1974).
6. R. G. Anderson, *Data Processing and Management Information Systems* (Macdonald & Evans, London, 1974).
7. G. A. Korn, *Minicomputers for Engineers and Scientists* (McGraw-Hill, Maidenhead, 1973).
8. D. R. Mace, T. Crowe and J. H. Jones, 'The Use of Minicomputers in Higher Education in the United Kingdom', *Int. J. math. Educ. Sci. Technol.*, 5 (1974) pp. 543—8.

3 COMPONENTS OF AN ELECTRONIC DIGITAL COMPUTER

Before we examine the components of an electronic digital computer let us consider how we ourselves would set about solving a problem. The functions of solving a problem can be broken down into a number of features.[1]

(a) Control – We must first of all plan and concentrate on what we must do; in other words control the way in which we are to solve the problem.

(b) Memory – The information we have stored in our brain will furnish certain facts, and our experience provides us with a means of solving the problem.

(c) Processing – From the information we have stored and that that we receive we must exercise our brain and process the data in accordance with the method we have decided to use.

(d) Input – Any information we receive is through our sense organs – our eyes, fingers, ears, nose and mouth. This then may be considered to be the way we receive data or the input to our problem.

(e) Output – Having worked out an answer, we would possibly record it for future reference. This is the output to our problem.

An electronic digital computer is composed of components that perform the same basic functions described above. These components are to be found in a *central processor* and *peripheral units* as follows.

Units in the Central Processor
(1) Control unit
(2) Memory unit
(3) Arithmetic unit (processing)

Peripheral Units
(4) Input devices
(5) Output devices

These components are known as the *hardware* of a computer (figure 3.1). We can liken the central processor to our own brain, which has the ability to 'control' actions, perform 'arithmetic' calculations or logic and 'memorise' facts. The units performing these functions are to be found in the central processor. The human brain has a limited capacity for memorising facts and information such as complicated formulae and other data has to be obtained from reference books. In the same way information is usually supplied to the computer from external sources. Data for the immediate problem to be solved are transferred to the memory unit of the central processor by devices not under the direct control of the computer. These are known as *peripheral units.* These units, when

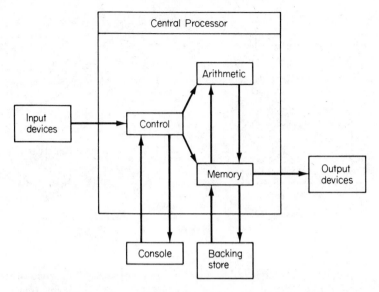

Figure 3.1   Components of an electronic digital computer

connected to the central processor, have an input or output function for putting data in or taking data out of the computer.

THE CENTRAL PROCESSOR

The main-frame component of the digital computer is the central processor, which houses the control, memory and arithmetic units.[2,3]

The Control Unit

An important part of the central processor is the control section, which is the master dispatching station and the clock of the computer. It directs the rhythmic flow of data through the system and controls the sequence of operations. To do this it interprets the coded instructions contained in the program and initiates the appropriate commands to the various sections of the computer. These successive commands, in the form of control signals, flow along circuits opening and closing switches and thus direct the flow of data in accordance with a program of instructions. The control unit automatically times and collates all the activities and ascertains that the computer is operating as a fully integrated system.

The Memory Unit

Another part of the central processor is the memory or storage unit.[4] This comprises a large number of character locations where data and instructions can be stored.

15

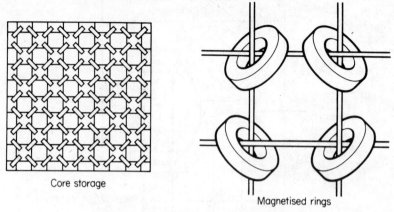

Core storage

Magnetised rings

Figure 3.2   Magnetic core storage

The size of the computer is described in terms of the size and speed of its memory and is defined by the number of characters it can hold. Information must be expressed in the form of binary digits, referred to as *bits*. A row of bits held in store is called a *word*. A word, then, is a set of characters which occupies a storage location. It is treated by the computer circuits as a unit and transported as such. A character location is called a *register*, which is found by a unique numerical reference called an *address*.

It is important to know which form of reference is implied since memory may be laid out as *characters*, *bytes*, or *word-orientated*, depending on the manufacturer's own concept of how the machine can best be used.[2] For example, in an ICL word-orientated machine a 'word' contains 24 bits and space is expressed in terms of 'K' (1024 bits). A computer having a size of 32K would have space for 32 768 bits.

The function of memory is to act as an immediate access storage. It is the size of this unit that limits the performance of the computer and it should be big enough to take program instructions, data to be processed, and the results of processing.

The unit comprises a magnetic core storage that has no moving parts (figure 3.2). It is made up of a square grid of wires similar to that in a pane of Georgian wired glass. Each intersection of wires is encircled by magnetised rings 1.5 mm in diameter. Pulses are sent down horizontal and vertical wires to change the magnetic state of any core from positive to negative, or vice versa. The magnetic state of each ring represents a binary digit. In practice each ring must be threaded with an extra wire to restore its original polarity and thus maintain its original state.

The Arithmetic Unit

The portion of the central processor that carries out the arithmetic and logical operations is called the arithmetic unit. The first thing to remember is that we are dealing with an electronic device, which means it has only two stable states,

16

such as on/off, current/no current and the like.[3] The following components generate conditions of a bistable nature.

(a) Pulses – Electric pulses, on a time scale, create a recognisable state of the presence or absence of a pulse.
(b) Switch – The position of a switch may be open or closed.
(c) Valve or transistor – The condition of a valve or transistor may be conducting or non-conducting.
(d) Voltage – A state may be registered by high voltage or low voltage.
(e) Magnet – Anything magnetised to saturation can represent a signal in one direction, and a second signal in the other direction. In the case of cores in memory the rings are electromagnetised and the polarity can be changed by electric pulses along the wires.

An 'on' state can represent 1 and an 'off' state can represent 0 in binary terms. The bistable nature of the electronic digital computer can be utilised to perform certain operations of arithmetic.

The binary system has two characters using 0 and 1. The x factor of 2 is represented by moving one place to the left. In binary four hundred and seventy-six would be represented as follows.

Factor	256	128	64	32	16	8	4	2	1
Number	1	1	1	0	1	1	1	0	0

Represents $256 + 128 + 64 + 0 + 16 + 8 + 4 + 0 + 0 = 476$

In contrast the decimal (denary) system has characters using figures from 0 to 9 inclusive. The x factor of ten is represented by moving one place to the left. In decimal terms four hundred and seventy-six is represented as follows.

Factor	100	10	1
Number	4	7	6

Represents $400 + 70 + 6 = 476$

Whatever the complexity of a mathematical problem it can be solved by the application of the basic rules of arithmetic – addition, subtraction, multiplication, division and exponentiation. The function of arithmetic is accomplished by internal processes that suit binary calculations.

The manual calculations involved in reducing a problem to these basic rules of arithmetic are too lengthy, and more advanced mathematical techniques must be used to speed things up. It is the speed of the computer and its ability to cope with a large amount of simple arithmetic in a short space of time that makes it particularly commendable.

Although the processing is performed in binary calculations the computer techniques will be better understood in decimal terms, as illustrated below.

Addition

This is simple and straightforward and needs no explanation.

Multiplication

This is accomplished by repeated addition.

Example
(i) 2 x 5 = 10 or 2 + 2 + 2 + 2 + 2 (five times)
(ii) 7 x 9 = 63 or 7 + 7 + 7 + 7 + 7 + 7 + 7 + 7 + 7 (nine times)

Division

This is accomplished by repeated subtraction.

Example
(iii) 24 ÷ 8 = 24 − 8 = 16 ⎱
             16 − 8 = 8 ⎬ subtracted three times
              8 − 8 = 0 ⎰

Thus
       24 divided by 8 = 3

Subtraction

This is accomplished by employing the process known as *complementary subtraction.*[2] The process might seem to be a round-about method and a little complicated. However, it is very useful since the complement of a binary number is expressed by complementing the digits and adding 1 to the result. The true complement of the binary figure range 101101 is therefore 010011 as indicated.

Number	101101
Complement digits	010010
Add 1	1
True complement	010011

The complement of a denary number is that number which must be added to it to give zero total. The decimal complement is those digits in the zeros concerned.

The process of complementary subtraction is as follows. If we wish to subtract B from A, which is $A - B$, add the decimal complement of B to A and disregard the most significant figure.

The decimal complement of 49 is 51, which is

     49 + 51 = 100

The decimal complement of 236 is 764, which is

     236 + 764 = 1000

The most significant figure in each case is 1.

Example (i)

     126 − 49 = x

Complementary subtraction

$x = 126 + 51$ (see above)
$x = 177$ less the most significant figure
$x = 77$

Therefore

$126 - 49 = 77$

Example (ii)

$496 - 236 = x$

Complementary subtraction

$x = 496 + 764$ (see above)
$x = 1260$ less the most significant figure
$x = 260$

Therefore

$496 - 236 = 260$

In decimal terms this method appears to be complicated and round-about. It is, however, quite a simple operation for the computer when one considers how the true complement of a binary figure is obtained.

BACKING STORE

Certain information can be held on storage devices as an extension of the computer's memory.[5] Information that is required for immediate access when solving a particular problem is held in the *core store* located in the central processor. Only a small proportion of information is required at any one time and it is practical to have the mass of data stored outside the central processor because of its limited storage capacity. Information needs may be stored either in the form of results of processing or for future reference on *backing storage*. The devices forming a backing store supplement the internal store of the computer. They are both input and output in the functions they perform. This is because information that is stored on the device may be fed into the computer as an input unit to assist processing. The results of the data processed by the computer may also be recorded on the device as output for future reference.

These devices constitute part of the memory of the computer and the selection of data is basically of two modes — *serial access* and *random access,* which is characterised by the way in which the data are located. Let us consider a memory with twenty-six character locations each of which has a unique address numbering 1 to 26. Each location contains a letter of the alphabet starting with 'A' in location number 1 and finishing with 'Z' in location number 26.

19

Problem   Find the words by using the contents of the following addresses in the order given.

    19,5,18,9,1,12   and   18,1,14,4,15,13

Process of Serial Access

To use this process we must start at memory address number 1 and proceed to memory address number 26 picking up the contents of each location on the way as they appear in the order in which they are referenced. On the first run-through the contents of memory address number 19 only can be picked up, which is 'S'. The second run-through will pick up the contents of memory addresses numbers 5 and 18. The contents of memory address number 9 will be picked up on the third run and the contents of addresses numbers 1 and 12 on the fourth run. This procedure must be carried out for every reference. Any references that are in a numerical sequence can be picked up on the same run-through. This process is synonymous with the preparation of a bill of quantities when billing direct from the take-off.

Process of Random Access

This process is much quicker since we can take a more direct approach in order to locate each character. The contents of memory address number 19 can be located first by making an immediate direct reference to that location. In the same way the contents of memory address number 5 can be located immediately afterwards. This procedure is continued until all the references have been made in the order in which they appear. Such references should yield the following information.

Address	19,5,18,9,1,12	18,1,14,4,15,13
		and
Contents	S E R I A L	R A N D O M

The backing storage devices are magnetic in character and the main types are a drum, tape or disc-type units. This method of storage is non-volatile. This means that information is not removed when power is switched off, as in the case of electric devices.

For long computations it is useful to store intermediate results from time to time, in the event of a fault. To switch off the power would interfere with the information on volatile storage devices. On these devices, after any fault has been cleared, the computation can restart from the last set of intermediate results. On volatile storage devices the processing would have to start right at the beginning.

Magnetic Drums[3]   This is one of the earlier types of storage unit, which comprises a heavy cylinder covered with a layer of magnetised sensitive material. This type of storage device is a random access store. The surface of the cylinder is divided into a number of narrow tracks on which data are recorded (figure 3.3). Each track has a reading and writing head associated with it. The cylinder

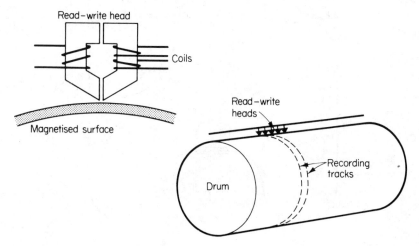

Figure 3.3   Backing store – magnetic drum

rotates at something between 2000 and 10 000 revolutions per minute and by careful timing the read/write heads can select the required 'bits'. If the drum is rotating at 3600 revolutions per minute each bit is positioned under the head 3600 times per minute or 60 times per second. The bit therefore can be stored or read in 0.0167 seconds (16 milliseconds). For a speed of 10 000 revolutions per minute the storing and reading time will be 6 milliseconds. On an average the input/output is in the region of 200K characters per second.

The main advantage of this type of storage device is its low cost. The main disadvantage is its comparatively slow speed: other devices work at speeds of fractions of a millisecond.

Magnetic Tape[2]   Information can be conveyed to and from the computer by means of a magnetic tape unit. This device is a serial access store. Data are stored on a continuous flexible linear recording medium impregnated or coated with magnetic sensitive material of high quality and held on reels (figure 3.4). The reels of tape are supplied in lengths from 365 m to 1100 m, and in widths from 12.5 mm to 25 mm. There are up to ten channels running side by side on a strip of tape on to which information is recorded. The density of recorded information is in the region of 250 to 360 characters per millimetre.

Although the magnetic tape units are permanently linked to the computer the reels are interchangeable. This means that a library of tapes can be built up and loaded on to the unit for processing by the computer as and when required. The tape units vary in speed according to the model and operate at between 10 000 and 160 000 characters per second. Generally the speed of a small to medium-sized computer would operate at a storing or reading time of 3 milliseconds per character, which is about 20 000 characters per second.

Information is set out on the surface of the tape in a sequential order in records. A record is the information about one particular item in the file. This

21

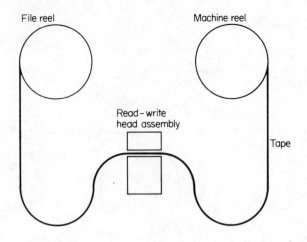

Figure 3.4   Backing store – magnetic tape

could be likened to the dimensions and description of an item of take-off. A number of records grouped together is called a 'block'. A block of information is that which is read for processing at any one time. Between the blocks of information is a gap known as an interblock gap where no recording is made. This is usually a standard 19 mm long. The object of this gap is to act as a brake to prevent the computer initiating further operations until the current information has been processed.

Magnetic Disc Packs[5]   Information is recorded on to the magnetised surface of a disc, something like a gramophone record, which is notionally divided into a number of concentric tracks. Each disc may contain 100 or 200 tracks or concentric rings that hold a fixed number of characters. A typical device consists of around six discs on a common spindle. This gives ten recording surfaces, since the top and bottom surfaces of the pack are not recorded (figure 3.5). This type of device is a random access store.

   All tracks in the same relative position on each of the ten recording surfaces are collectively known as a cylinder. A disc pack therefore has a number of concentric cylinders of data. The discs are very often interchangeable, which offers great flexibility in their use, although some computer users have disc devices that are fixed.

   The pack shown holds about 9.2 million characters and rotates at a speed of 2400 revolutions per minute. This means that information is transferred at around 200 000 characters per second or 25 milliseconds per character. The seek time varies between 30 and 160 milliseconds.

   The device is useful where very large files are to be held 'on line' to the computer. In general they are too large and expensive for average commercial use. They do, however, have a very large data-holding capacity of something in the region of 300 million characters or more.

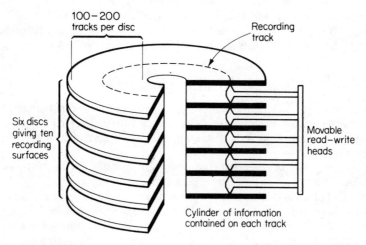

Figure 3.5 Backing store – magnetic disc pack

On-line and Delay-line Storage

The peripheral equipment or devices in a computer system that are under the control of the central processor are said to be 'on line'.

The availability of data and their immediate processing may be a problem in some systems. Information may be made available at a time when the computer is not quite ready to process it — how often does one make an appointment with a doctor or a dentist, and turn up on time only to find him delayed by previous appointments? The facility for keeping data circulating in the form of pulses in specially designed circuits is called *delay-line* storage.[3] This form of storage is not permanent since when the circuit is switched off the pulses cease. The storage techniques allow the data to travel through some medium such as mercury, which causes a reduction and distortion in the pulses. As the delayed pulse emerges from the delay-line medium it is passed through an amplifier and is reshaped and enters the system as a delayed pulse for processing (figure 3.6).

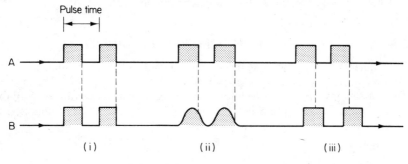

Figure 3.6 Delay-line storage

23

(i) Pulses A and B are identical in timing.
(ii) Pulse B passes through the delay-line medium, is reduced and distorted.
(iii) Delayed pulse B emerges and is amplified and reshaped.

COMPUTER CONFIGURATION

An electronic digital computer is made up of several devices that basically serve one of five functions. Each type of device has its own characteristics, which may differ slightly from a similar device serving the same function, as outlined earlier in this chapter.

Interconnections between the computing units are made with shafts, wires or even pipes, depending on how the units work. The arrangement of interconnected units is known as the 'configuration' and the preparation of pieces of equipment for operation to solve any particular problem is known as the *set-up*. We may think of a computer installation as containing a stock of devices of each type, from which sufficient numbers are drawn to solve any particular problem. The configuration for solving a complex problem may be dealt with in one of two ways, by arranging the units to act either simultaneously or sequentially.[3,6]

Either: different units are interconnected, each being programmed to process a particular, and therefore a simplified aspect of the problem. The coalescence of results by the simultaneous actions of the device produces a solution. A computer with this type of set-up is known as a *simultaneous computer*. It is a characteristic of a simultaneous computer that several specimens of a particular type of computing unit are needed, in fact as many as the number of times that the problem calls for a particular computing operation. A further characteristic is that a given interconnecting path in the set-up carries signals representing successive values of the same operation. It never carries values of different operations.

Alternatively: the problem may be broken down into simplified stages of processing. Each stage is processed by the same units performing different operations. For instance, one arithmetic unit might be provided to solve all the various calculations needed in the solution of the problem. Interconnecting paths in the set-up carry signals of different operations at different times, for each stage of the processing. A computer with this type of set-up is known as a *sequential computer* — this is the more common approach.

COMPUTER PROCESSING TECHNIQUES

The problem of data processing by computer may be solved in a number of different ways depending on the means of access to the computer.[5] The methods of processing involve various techniques, the uses of which are influenced by the hardware and software available. The range of techniques may be summarised as follows.

(a) Batch processing
(b) Real time processing
(c) On-line processing

(d) Off-line processing
(e) Multiprogramming
(f) Simultaneity

(*a*) *Batch Processing*   involves the grouping of related data for processing. The data are processed in a predetermined sequential order in groups called batches. This means that the data are collected in batches and processed by using programs to solve a particular problem to which the data relate. Each computer run may involve the use of files that may be on magnetic tapes or discs and that are removed after the processing has been completed. It is not possible to gain access to any record until the relevant file has been specially set up for the purpose. If the data are required to be processed and files interrogated as events occur other facilities are necessary.

(*b*) *Real Time Processing*   is a non-sequential method designed to meet certain information demands requiring immediate processing, and is concerned with computer communications involving the interrogation of data on files. For instance, a manufacturer may wish to know the state of his stock, or an airline the availability of seats on a particular flight; banks use this method of processing to service enquiries regarding bank statements and the like. The system satisfies the need of many businesses to gain access to current information. Processing takes place in the time available to make decisions that affect events and involves a 'while you wait' situation. A more complex configuration is required for real time processing, which creates a need for random access storage and on-line equipment (see on-line processing). A powerful centrally located computer with a high capacity direct access storage in the form of fixed disc units is required. This type of processing is expensive and because of its complexity certain problems are created regarding reliability and recovery from error.

(*c*) *On-line Processing*   relates to those operations performed by devices connected to and under the direct control of the central processor. Batch processing is normally associated with hardware that is directly connected to the central processor and is located in the computer room. Peripheral equipment that is on line means those devices that are connected to the central processor for the purpose of processing. There are a number of devices that can be connected to the computer over the normal Post Office and telegraph lines or private lines. These enable the computer facilities to be made available at almost any distance from the central processor itself — they are known as communication terminals.
    The more common devices are as follows.

(i) The ordinary G.P.O. telex machine is used over a dialled telegraph network normally from one telex machine to another. The operator simply dials the number of the computer and transfers the data via a keyboard. Printed answers are received from the computer.
(ii) A variation on the same principle, but designed for use over private lines, is the much cheaper teletype machine. This is just as effective as the telex machine and can also be used for transmitting via paper tape into the

computer and can receive punched paper tape as an output in addition to printed output.
(iii) Another device known as the visual information projection device or visual display unit (V.D.U.) has a cathode-ray tube very much like the ordinary television set but it is equipped with a keyboard. The operator can see on a screen what is being transmitted. Information can be received back visually in readable form on the screen.

Although on-line processing relates to those operations performed by machines directly connected to the central processor the term is usually associated with a method of processing that uses a terminal at a remote location. The operations are usually performed in real time. However, it is possible to use on-line equipment for batch processing. Data keyed in at the keyboard at a remote location can be entered either directly into the computer without any further human intervention, or recorded on paper or magnetic tape and stored for processing at a later time. This method is known as remote batch processing or remote job entry.

(*d*) *Off-line Processing*   is carried out by machines that are not under the direct control of the central processor and the processing is not directly associated with the operations of the central processing unit. Human intervention and control are required between data entry and ultimate processing. Certain data-processing tasks may be performed by auxiliary equipment although this is relatively slow compared with the operations of the central processing unit. The following examples illustrate some of the off-line operations that may be suitably performed by this type of configuration.

(i) Cards may be sorted into a master file sequence when it is not required to sort the data by a computer program. The operation would be carried out by a mechanical sorting machine.
(ii) Data may be transferred from one media to another in order to facilitate a speedier transfer of the data to the computer memory. For example, data on punched cards may be transferred to magnetic tape. The operation would be carried out by a card reader and a control unit connected to a magnetic tape unit.
(iii) Data held on magnetic tape may be transposed into a printed report. This is a much more economic operation than that through the central processing unit which involves the transfer of the data to core store and their subsequent print-out using an on-line printer.

These tasks may be more suited to off-line processing thereby enabling the computer to be more fully utilised by relieving it of the slower time-consuming operations.

(*e*) *Multiprogramming*   is a technique involving the interleaving of two or more different programs for simultaneous processing. Computer power may be made available on a wider basis enabling users to gain access by means of data transmission services. Real time computer systems may provide facilities simultaneously, which serve a number of users with randomly occurring requests. The system may be used to serve multiple organisations with many users in the same organisation sharing the same machine or it may serve many smaller users.

A coordinating supervisor program is required to control the interchange between programs. This usually forms part of the executive routine and on large computers is held in memory ready for processing. Only one program may be executed at any one time. The operations are controlled to allow programs to be switched. For instance, if processing is interrupted on a program run by the need to attend to a peripheral operation, the central processing unit will switch to another program in the meantime. A situation is created whereby one program is always active with other programs either awaiting the completion of peripheral operations or awaiting the availability of the central processor. This method of using the computer on a time-sharing basis is known as multiprogramming.

(*f*) *Simultaneity*   is a technique involving the overlap of the input and output operations with the calculating processes to form concurrent operations. In a number of instances simple calculations may be required on a large quantity of basic data. The central processing unit is involved for a very small proportion of the processing time since each calculation is simple and the speed of processing is more directly related to the input and output operations. If the central processor is rendered idle during the relatively slow operations performed by the input/output devices the advantage of working at electronic speed is not achieved. This situation may be avoided if the central processor and its peripherals work concurrently. For example, the peripherals signal the central processor whenever they are available for data transfer. Basic calculations performed by the central processing unit may be interrupted to enable it to deal with the peripherals and may be renewed after attending to their requirements. By this method the central processor and its peripherals work concurrently.

INPUT DEVICES FOR NEW DATA ENTRY

The computer is not able to receive data in their original form; they must be converted into a form that is recorded on a medium recognised by the machine. Since the computer is in a bistable state when active, the data that are fed into it must also be of a bistable character. The following is a summary of the possible media that the computer will recognise.

Punched Cards[2,6]   Information is punched in a coded form on cards containing forty, sixty-five or eighty vertical columns. Each column has twelve possible positions in which a hole or holes may be punched (figure 3.7, which shows one of three national codes in use). Ten of the positions represent the numbers 0 to 9 and additional positions provide combinations which form other characters.

The peripheral device used on line with the computer for introducing information on punched cards is the 'card reader'. This device senses the pattern of holes in each column of the card and converts them into electric pulses. Data are read into the computer at an average speed of 800 cards per minute although devices can be used at speeds ranging between 400 and 1200 cards per minute. It

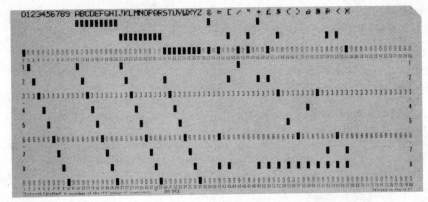

Figure 3.7   Punched card

is difficult to assess the actual amount of input since the number of characters per card will vary. The average speed of reading is about 900 to 1000 characters per second.

In order to read a punched card in alphanumeric terms it must be passed through a device called an 'interpreter'. This machine senses the pattern of holes and prints the characters at the top of the card.

Copies of punched cards may be obtained by using a device called a 'reproducer'. This machine is useful where extra copies are required or where old cards have become worn and unserviceable. One advantage of this medium is its flexibility: cards can be renewed and replaced in any sequence. This leads to one big drawback since it is easy for cards to go astray.

Punched Paper Tape[2,6]   Information is punched in a coded form on to a continuous strip of paper tape in rows of five, six, seven or eight (figure 3.8). Each character is formed by holes punched across the width of the tape. The device used for paper tape input is a 'paper tape reader', which detects the pattern of holes in a similar manner but a little faster than the punch card reader. This type of medium has the advantage of having a variable field length and therefore is more flexible than the punched card, which has a fixed field length of eighty columns.

Both punched cards and punched paper tape should be completely accurate, otherwise results would be unreliable and useless. The process for ensuring the correct transcription from the source documents to this medium is called 'verification'. This process is performed by a second operator punching the same information to enable any discrepancies to be detected when the relative positions of the holes are compared.

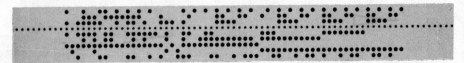

Figure 3.8   Punched paper tape

28

Other Forms of Input Media

The following is a brief summary of possible media in addition to punched card or paper tape.

Magnetic Ink Characters[5]   This medium is used by banking systems and consists of stylised types of numeric characters such as those found at the bottom of cheques. The characters are formed with ink containing particles of a magnetic substance that can be detected by an automatic device known as 'Magnetic Ink Character Recognition' (M.I.C.R.) – up to 1600 cheques per minute can be read.

Optical Readable Characters[5]   A form of optical reader can be used to read the special shape of each character printed on the input media. The device translates the data into electric pulses, which in turn are transmitted to the computer for processing.

Visual Display Unit (V.D.U.)[7]   Data as they are input are displayed on a cathode-ray screen like a television screen. This particular form of input display is used for medical record schemes and architects' computer-aided design.

Communication Terminals[8]   Certain devices known as 'communication terminals' can be connected to the computer over the normal Post Office or telegraph lines. These enable facilities to be available at almost any distance from the central processor itself. By the use of a remote data communication terminal data keyed in at the keyboard at a remote station can be entered either directly into the computer, without any further human intervention, or transferred to paper or magnetic tapes as delay-line storage. This type of device can be extremely useful if installed in a quantity surveyor's office, enabling him to have a direct contact with a computer (see chapter 11).

OUTPUT DEVICES FOR RESULTS

The results of processing data are 'output' from the computer. This output may vary in its form and may appear as characters printed on sheets of paper or coded on punched paper tape or card. Devices that make the results of processing available in a form easily read are different types of printer. These may take on any one of the following characteristics.[2,3,6]

Single-character Printers   This type of machine prints single characters at a time in a similar manner to a typewriter. The maximum output is approximately ten characters per second.

Line Printers   The feature of this type of machine is its ability to print whole lines of 120 character positions simultaneously. This is the most common form of legible output. There are a number of different types of machine each having its own printing action. Some machines are faster than others and output varies between speeds of 300 to over 2000 lines per minute. The normal speed is about 650 lines per minute.

Optical Printers   This type of printer uses a process that has similarities to the 'xerographic' method of document reproduction. A cathode-ray tube is used to project a line of print on to a revolving drum with a light-sensitive surface. The images so produced are transferred to paper by means of a powder dusted on to the drum and fixed to give a permanent image by a heat process. The advantage of this type of printer is its ability to reproduce any shape, and it is not limited to the fixed set of characters found on normal printers. It can be used to print diagrams and charts or any image that the cathode-ray tube can produce, and is useful for printing architects' drawings from a computer-aided design system. Use is also made of laser beams and graph plotters.

Console Typewriter   A device used to monitor the progress of computer operations is the console. It is not often thought of as a true peripheral device but is in fact a means by which the operator communicates with the computer by keying in instructions on the keyboard. The device might be programmed to request information from the operator or to report the progress of operations back to the operator by typing out messages. It can also be used as an auxiliary printer for short reports to the management and thus it is truly an input and output peripheral device.

Coded Output   Other types of output device may produce media similar to the input media already mentioned, such as punched cards or punched paper tape.

In the earlier types of calculator, numerical information was punched on to an input card for feeding into the computer. After processing, the card emerged with added punched information as output. Nowadays separate cards are punched as output. As an output media punched cards are not very popular because of the relative speed of punching, which varies between 100 and 400 cards per minute.

In certain instances where a small computer has a limited storage punched cards may be the only alternative means of storing data for processing. There are disadvantages in using cards in this way since there is a limited amount of data (a maximum of eighty characters) that can be stored on any one card, therefore a large number of cards would be required for the mass of data that need to be stored.

The production of punched paper tape as an output medium operates in a similar fashion to that in the input section, with the exception of the reader, which is replaced by a punch. This method of output is used as a means of creating data to be subsequently transmitted directly over telegraph or telephone lines from communication terminals to the central processor.

REFERENCES

1. A. Vorwald and F. Clark, *Computers from Sand Table to Electronic Brain* (Lutterworth, Guildford, 1966).
2. T. F. Fry, *Computer Appreciation* (Butterworth, London, 1970).
3. H. Jacobowitz, *Electronic Computers Made Easy* (W. H. Allen, London, 1967).
4. E. M. Awad, *Automatic Data Processing* (Prentice-Hall, Englewood Cliffs, N.J., 1970).

5. R. G. Anderson, *Data Processing and Management Information Systems* (MacDonald & Evans, London, 1974).
6. S. H. Hollingdale and G. C. Tootill, *Electronic Computers* (Penguin, Harmondsworth, 1966).
7. B. Auger, *The Architect and the Computer* (Pall Mall, London, 1972).
8. ——'Data Transmission', *The Computer Users' Yearbook*, ed. P. Grant (Computer Users' Yearbook, Brighton, 1974).

4 OPERATING THE COMPUTER

It must first be appreciated that the computer does not and cannot solve problems — it merely carries out instructions. If the solution to any problem is achieved by blindly carrying out these instructions, so much the better. In order then to make the computer do its work there must be a set of instructions prepared for the purpose of solving a problem. These instructions are fed into the machine to direct the automatic processing. A plan for the solution of a problem must include procedures telling the computer exactly how to handle each incoming bundle of facts, how to process them and what to do with the results after they have been processed. This procedure is known as *programming*, and the person who prepares the instructions is known as a *programmer*.

The computer cannot think or even take decisions, although it may appear to do so on occasions. In reality alternatives have to be seen by the programmer. Any long involved calculation or routine might proceed in one of a number of directions depending on some intermediate result. The programmer may not see the result but can foresee the possibilities. It is important that all contingencies be covered, since the computer is incapable of even the smallest extension to its instructions. Any unforeseen situation that has not been covered by any instructions will cause the machine to either stop its run or impart spurious results. All problems must be closely analysed and the procedure clearly and logically set out. Programming, then, is the exercise of a rational thought process. It is more of an art than a science since there may be a multiplicity of ways of instructing the computer to solve a problem. No two programmers are likely to come up with exactly the same program.

The technique of programming, like the preparation of a bill of quantities, is broken down into a sequence of logical operations. The first major steps to be taken involve (a) the analysis of the problem and (b) flowcharting as an aid in establishing the facts and rationalising a method of approach (figure 4.1).

ANALYSIS

The first step is really one of defining the logic. Given a problem, the programmer will set it out in precise terms and decide the general way in which he proposes to obtain a solution. In order to do this the main activities must be established and put in a logical sequence. These must be broken down into more simple steps that comprise a series of procedures each forming a complete cycle. For example, the main activities necessary for the preparation of a bill of quantities are as follows.

(a) Taking-off
(b) Squaring and casting dimensions
(c) Abstracting
(d) Billing
(e) Editing

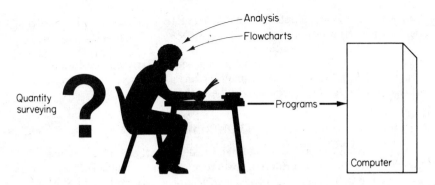

Figure 4.1   Programming routine

Each activity involves a logical sequence of operations that must be examined and analysed in detail with the object of integrating the manual processes with the computer requirements. It is important that the current procedures used for solving a problem be examined, and it must be considered whether any significant contribution can be made by the computer in time saved, efficiency and cost.

Taking-off   Taking-off is effected basically in the format used for booking down the dimensions. A facility that the computer can offer is its ability to store information. This enables the techniques of short coding (see chapter 8) and unit quantities (see chapter 9) to be used to advantage as aids to speeding up the processes of taking-off and coding.

Squaring   While the computer has the means for squaring dimensions it is not very often used for this purpose. The problem lies with the input media, since each dimension has to be reproduced in code and this is time-consuming. To use the computer as a calculator may be considered by some to be uneconomic.

Abstracting   The main problem in the production of a bill of quantities lies with the method of processing. It is in this activity that the computer can be of the greatest help. A number of solutions have been developed, each as a result of the need to solve a particular situation.

Billing   The problem of billing is a matter of output from the computer. Once the data have been processed there seems little difficulty in programming the computer to produce a print-out with a satisfactory format. If the program is correct a draft bill can be produced in almost any format, such as sectional, elemental, operational, and so on.

Editing   This is still a manual process in the sense that there are certain mistakes that can only be found by manual inspection. Generally a program will contain a routine that will give a print-out for scrutiny after the process of

33

abstracting has been completed. Any errors that are detected can be corrected before the draft bill is printed.

The development of routines necessary to solve a problem does not fall into place automatically and much thought is needed to arrive at a satisfactory program. A suitable means of rationalising one's thoughts is to use some form of visual aid; this can be achieved by the use of flowcharts.[1]

FLOWCHARTS

Having analysed the problem and established the method and procedures by which a solution can be achieved, some record must be made. The usual method of doing this is to prepare a *flowchart* with appropriate explanations. This is a graphical representation of the sequence of operations with the use of symbols. Apart from being a convenient way of recording a method of arriving at a solution, flowcharting serves a number of useful purposes, as follows.[2]

(a) It is a valuable aid in organising the analytical thought process.
(b) The format of a flowchart helps to split up the problem into parts that are more easily understood.
(c) It is a means of communicating thoughts to others.
(d) It is a means of memorising a method that can be easily recalled.
(e) Quite often a number of possibilities have to be considered and methods of achieving a solution involve trial and error. Empirical attempts at finding a method can be amended easily.

The first thing to do when constructing a flowchart is to make a note of the aim of the chart. This will have an influence on the method of approach. There are a number of types of chart, each having a different aim, namely the *block chart, systems flowchart* and *program flowchart*.[3] Each one deals with a particular aspect of the method of achieving a result.

Block Chart   This is the simplest form of flowchart, the primary object of which is to indicate the main sequence of operations within a system without showing, in detail, how they are carried out. The block chart uses only one symbol — a rectangle into which a short description of the operation is inserted. The interrelationship of operations is indicated by flowlines with arrows showing the direction of flow. A simple block chart, showing the operations involved in the preparation of a bill of quantities, can be seen in figure 4.2. This type of chart is essential in rationalising the operations and establishing the procedures to be performed. It might be found necessary to perform certain operations manually and others by the computer.

Systems Flowchart   The systems flowchart is concerned with the system as a whole and deals more specifically with the manual and computer processing operations. A little more detail of the routine may be shown on this type of chart, using standard flowchart symbols[4,5] (see figure 4.3). Each symbol is accompanied by a short description of the operation, which is either written in the boxes, or appended separately, but referenced to the boxes. Often it is necessary to prepare a series of charts for different sections of the routine.

34

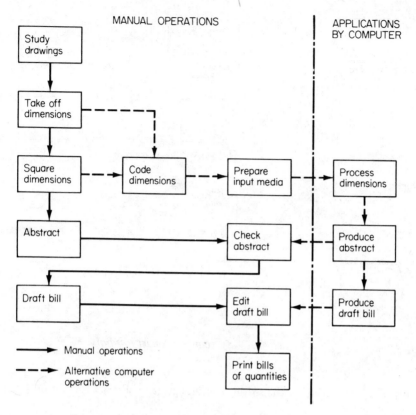

Figure 4.2   Block chart

These are interrelated by *connector* symbols. This is because the additional detail makes it impractical to show the whole routine on one chart. The routine might conveniently be split into such sections as: manual procedures, data preparation, computer processing and methods of dealing with output data. A simple flowchart showing certain routines for the preparation of bills of quantities is shown in figure 4.4. This analyses in a little more detail the computer applications for the system established in figure 4.2. The flowchart indicates the type of input and backing stores. In this instance the chart shows the input media to be punched card or punched paper tape and the backing stores to be magnetic tape or disc. The flowchart also sets out certain problems that need to be resolved in the implementation of the system. The auxiliary operations indicate the objectives and the various programs to be written. For instance the programs required may be summarised in general as follows (see figure 4.4).

(a) Extract the items from the input media, check for errors and record the data.

(b) Sort the data as required by the abstracting process and record the results.

35

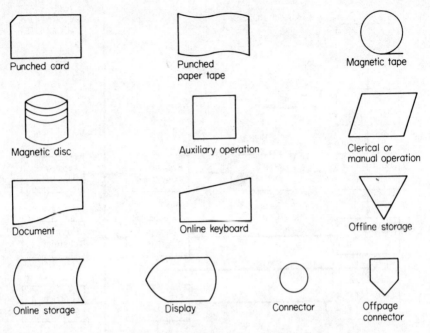

Figure 4.3   Systems flowchart symbols

(c) From the results of the sorting routines print an abstract for editing purposes.
(d) Transpose the abstract data into a format required for the bill of quantities.
(e) Print a draft bill of quantities.

The foregoing indicate the objectives of the various programs that need to be written. Just how these objectives are achieved depends on the skill and knowledge of the programmer who will have to produce a program flowchart before he can consider writing any program.

Program Flowcharts   A program flowchart deals with the part of the routine that specifies the computer instructions.[2,3,5] This is concerned with a clear definition of a set of instructions telling the computer how to handle the problem. Flowchart symbols are shown in figure 4.5. This process must be carefully considered so that all the necessary details are provided to deal with possible alternatives. Great care must be exercised in formulating the instructions. Although the computer might appear to think and take decisions it is the programmer who has seen the alternatives and instructed the machine accordingly. Programming is a matter of trial and error and many attempts will be made, with modifications, before a satisfactory solution is reached. The solutions to complex problems might proceed in a number of directions depending on some intermediate result. The programmer cannot see the result but can foresee the possibilities. A simple example is that of telling the computer

36

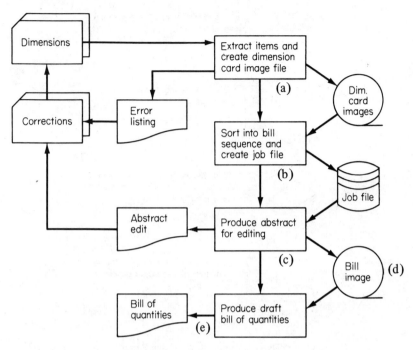

Figure 4.4   Systems flowchart

how to recognise positive and negative dimensions and to select the appropriate section for processing.

An example of a program flowchart (figure 4.6) indicates a routine for calculating the total wall area of a number of rooms and printing the answer. The procedure is as follows (see figure 4.6).[5,6]

(i)   A 'start' instruction is given to set the computer working.
(ii)   The computer is told which equipment to use and to reserve the relevant amount of space for setting up the files and for due processing.
(iii)   The computer is told to read a card and store the information.
(iv)   At this stage the computer may find that the last card has been reached, as indicated by an end-of-file marker which may be (****).
(v)   Assuming that a card has been read, the data can be processed and the room area calculated (see chapter 5).
(vi)   Each room area is added to the total area.
(vii)   Another card is read, the data processed and the room area added to the total area. This procedure is repeated until the last card is reached as indicated by an end-of-file marker. This procedure is known as *looping*.
(viii) All the results that are required to be printed are transferred to a *print file*, which sets out the format to be reproduced.
(ix)   Having collected all the information and assembled it into the specified format, the computer is instructed to print it as output.

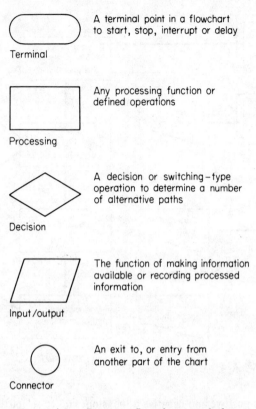

Terminal — A terminal point in a flowchart to start, stop, interrupt or delay

Processing — Any processing function or defined operations

Decision — A decision or switching–type operation to determine a number of alternative paths

Input /output — The function of making information available or recording processed information

Connector — An exit to, or entry from another part of the chart

Figure 4.5   Program flowchart symbols

(x) On completion of the output process the files can be closed.
(xi) The last instruction is 'stop running'.

The procedure of flowcharting is a kind of filtration process in the analysis of problem situations and is performed by a number of individuals including systems analysts and programmers (see chapter 11). The final result is a program specification. A problem may be segmented by defining broad ideas that are expanded and synthesised into a functional working system as established by the block chart. From the block chart will emerge computerised procedures that can be summarised on a systems flowchart.

The systems flowchart should be drawn up in sufficient detail to be able to recognise the need for individual programs. To obtain the right program it is essential that a good program flowchart be designed before writing a series of instructions. It should be appreciated that the program flowchart is not intended to be a detailed description of every program step or command. Each symbol represents a step in a logical sequence. The program flowchart forms a blueprint for use by the programmer. Large complex programs may require a program flowchart that initially sets out the main line logic. From this chart large

38

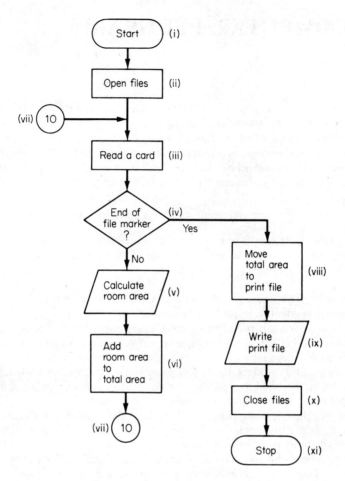

Figure 4.6   Program flowchart

segments will be extracted and described in more detailed supplementary program flowcharts. This technique is known as *modular program flowcharting*. Many errors are due to faulty logic and a program flowchart is very useful when the program is being tested.

REFERENCES

1. J. Maynard, *Computer Programming Made Simple* (W. H. Allen, London, 1972).
2. M. Bohl, *Information Processing* (Science Research Associates, Henley-on-Thames, 1971).
3. T. F. Fry, *Computer Appreciation* (Butterworth, London, 1970).
4. BS 4058 : 1973 Data processing flow chart symbols, rules and conventions.
5. D. K. Carver, *Introduction to Data Processing* (Wiley, Chichester, 1974).
6. J. Watters, *COBOL Programming* (Heinemann, London, 1972).

5 COMPUTER PROGRAMS

Software is a term that is complementary to *hardware* and refers to all the programs used by the computer. The types of program used to operate a computer may be classified as (a) operating-system programs and (b) user-written programs. Operating-system programs are required to regulate the computer operations and are supplied by computer manufacturers. User-written programs are used in computer processing for solving individual problems.

OPERATING-SYSTEM PROGRAMS

The operating-system programs govern the techniques and procedures for operating the computer. The most important of these programs are the *supervisor* and *assemblers* or *compilers*.

Supervisor[1]   This is a program supplied by the manufacturers and is a permanently resident program held by the computer in core storage while the computer is working. It performs an executive routine, which controls the operations of other routines and programs and is sometimes known as the 'executive'. The function of the supervisor is to control the automatic operations of the computer by reading, interpreting and actioning the commands supplied to the system. Its function might involve the control and sequencing of many programs at the same time.

Another function of the supervisor is to keep the computer operators informed of the current state of the computer's operations. Messages will be transmitted automatically through the computer operator's console typewriter indicating such problems as records that cannot be found, or indicating unsuccessful input/output operations. The computer operator can also request information by typing the message on the console for transmission to the computer. This will be read by the supervisor, which will collect the relevant information and cause it to be printed on the console.

Assemblers and Compilers[2,3]   The computer is a machine that is in a bistable state and functions in a machine language comprising numeric binary codes. This type of code is made up of digits that convey the data and instructions. The digits are formed by a binary character representation as described in chapter 3. How, then, can the computer operate if instructions are prepared using a program written in a language containing characters used every day for human communications? The problem is one of translation. The translation of programs into the computer's own machine language is achieved with the aid of special programs written for the purpose. These are stored in the central processor as *assembler* programs or *compiling* routines.

A program written by a programmer using alphabetic and numeric characters is known as a *source* program.[3] The machine instructions developed by the translating routines of the assemblers or compilers form a program in a machine language that is known as an *object* program. The source program is read in as data and instructions and the assembler or compiler governs the conversion of the computer language statements into the machine language instructions of the object program. The results of this process (that is, the object programs) are stored on a backing store such as magnetic tape or disc.

An *assembler* program is one that operates on user-written programs and produces machine instructions. It carries out such functions as translation of symbolic codes into operating instructions. The translation is performed item for item and produces the same number of instructions as defined by a user-written program.

A *compiler* is a program like an assembler program that is written to perform a translating routine. A compiler is more powerful than an assembler, for in addition to its translating function it is able to replace certain items in a user-written program with a series of instructions. While the assembler translates item for item, the machine instructions that result from a compiler are an expanded version of the original. This means that it is able to generate a number of instructions in the object program from a single instruction in the source program.

USER-WRITTEN PROGRAMS

The detailed instructions contained in a program flowchart must be transposed into a computer language. This is an artificial language for describing or expressing the instructions that can be carried out by a digital computer. The machine language takes the form of numeric binary codes that are peculiar to each type of machine. It is almost impossible to write instructions in machine code terms and therefore a symbolic language is used. Various languages have been developed which may be described as *low-level* or *high-level* languages depending on the extent to which they are machine-dependent or problem-orientated.[2]

A Low-level Language   is one that uses a mnemonic code and is machine-dependent and bears a close resemblance to the machine code. It has a one-to-one relationship between the written instruction and the machine instruction and is assembled into the computer by an assembler program. To write a program in a low-level language requires a detailed knowledge of the type of computer for which the program is being written. In this situation programming is a specialised operation performed by highly trained personnel. Any change of computer type would necessitate an updating in the education of the programmer.

A High-level Language   has been developed which, from the programmer's point of view, has made programming simpler. The programmer is freed from the constraints imposed by any particular computer he happens to be using and he is able to write his programs in the way that is most convenient to him. Instructions may be written in ordinary English phrases that will generate a number of machine instructions.

High-level languages are machine-independent and have a one-to-many

41

relationship between the written instructions and machine instructions. A program written in a high-level language is problem-orientated rather than machine-dependent. This means that the languages have been developed for use in dealing with specific types of problems rather than catering for the idiosyncrasies of the various computers. A program written in a high-level language can therefore be placed on any type of machine that has a suitable compiler stored in its central processor. It is the compiling routine that translates the program into machine instructions in machine code terms. This type of language may be considered to fall into certain classifications: (a) those with scientific and engineering applications, (b) those with business and commercial applications and (c) interactive (conversational) computing.

Languages for Science and Mathematics

The languages possessing qualities for processing scientific and engineering data are concerned with a numeric approach and have a mathematical bias. These may relate to such problems as $y = 3x + 2$. Some of the better-known languages for these applications are as follows.

ALGOL (ALGOrithmic Language[2,4])　An algorithm describes a step-by-step procedure for solving a complex problem. The forerunner of ALGOL was the ALGebraic Oriented Language. ALGOL has the advantage of having international recognition as a common language. The language itself is a result of international cooperation to obtain a standardised algorithmic language. A program written in this language presents numerical procedures to a computer in a standard form.

FORTRAN (FORmula TRANslator[2,5])　This language was developed by IBM and is machine-independent. It is procedure-orientated, which means that it is acceptable by any computer that has a FORTRAN compiler. The problem-solving procedure is specified by a series of English—mathematics statements. Since the language is very similar to normal scientific and mathematical usage it can be written fairly easily by the mathematician or scientist without him having to be a computer expert.

Languages for Business and Commercial Applications

A well-known language used for commercial applications and suitable for quantity surveyors' processing is COBOL.

COBOL (COmmon Business Oriented Language[6,7])　This is a problem-orientated language that was developed for commercial usage and also has international recognition; as such it can be used for quantity surveyors' programs. It can be used on any computer that contains a suitable compiler and a central processor with sufficient core storage capacity. The language statements contain basic key words and symbols that specify a definite set of machine instructions. The programmer can write his program easily, rapidly and accurately using English words and conventional arithmetic symbols. The program is written in four divisions, as follows.

(a) Identification division — this contains all the documentary information relating to the program, such as the name of the program and programmer, the date written and any other relevant information.
(b) Environment division — this division specifies the hardware on which the COBOL compilation is to be accomplished.
(c) Data division — details are given of the data that are to be processed.
(d) Procedure division — the logical steps for the due processing of the data are outlined in this division.

An example of COBOL instructions in the procedure division of a program is as follows.

```
PROCEDURE DIVISION.
START.
AREA-CALCULATION.
       READ CARD-FILE AT END GO TO PRINT-OUT.
       ADD LENGTH WIDTH GIVING SUM.
       MULTIPLY 2 BY SUM GIVING PERIMETER.
       MULTIPLY HEIGHT BY PERIMETER GIVING ROOM- AREA
       ADD ROOM-AREA TO TOTAL-AREA.
       ADD DOOR-AREA WINDOW-AREA GIVING OPENING-AREA.
       SUBTRACT OPENING-AREA FROM TOTAL-AREA.
       GO TO AREA-CALCULATION.
PRINT-OUT.
       MOVE AREA-CALCULATION TO PRINT-FILE.
       WRITE PRINT-FILE.
       CLOSE CARD-FILE PRINT-FILE.
       STOP RUN.
```

This example illustrates a portion of a COBOL program for the calculation of the total net wall area for a number of rooms, as required for the problem set out in the program flowchart (figure 4.6) in chapter 4.

AREA-CALCULATION is a sequence in the program that provides instructions for the calculation of the net wall area of a room and carrying forward the totals. PRINT-OUT is a sequence for collecting the data and printing.

The data supplied consist of the following.

 (i) The length of a room
 (ii) The width of a room
(iii) The height of a room
(iv) Details of door openings
 (v) Details of window openings.

The calculations of the door and window openings are produced elsewhere in the program in the computer storage locations indicated as DOOR-AREA and WINDOW-AREA. Storage areas in the computer for the area calculations are

43

indicated as SUM, PERIMETER, ROOM-AREA, OPENING-AREA and TOTAL-AREA as set out in the data division of the program.

Languages for Interactive (Conversational) Computing

A programming language that was primarily designed for interactive (conversational) computing is called BASIC. BASIC was originally created for use in a time-sharing computer system developed at Dartmouth College in 1959.[8] It is used as a means of communication over a data terminal that is on line with the computer. Like other languages BASIC has to be translated from a source program to an object program although the user is unaware of the processing. The compilation and execution of the program are consecutive steps in processing. The system permits the user to enter his program and issue commands through the data terminal, which is in direct contact with the computer. This means that processing is carried out while the user is on line. Messages such as errors or requests for further information from the computer will be relayed back and forth between user and computer.

Any problems that need to be solved in an interactive situation can suitably be processed by using a BASIC program over a data terminal. The user can write his program in simple terms step by step and get an immediate response to his problem.

FORTRAN is another language that may be used in an interactive situation provided that the computer is supplied with a modified FORTRAN compiler. This system of computing involves the concurrent use of a computer by a number of individual users at a terminal some distance from the central processor, each using a different program. The technique requires the interleaving of these programs to enable the computer to select the appropriate program as and when it is required. This is known as *multiprogramming*. The system requires a coordinating supervisor program to control the switching of one program to another.

COMPILING AND TESTING PROGRAMS

Before a program may be used for processing it must be suitably compiled and tested.[2] These processes are illustrated in figure 5.1. As each source program instruction is read into the computer it is scanned by the compiler for errors in syntax — these are errors in the construction of the program statements only, since the routine is unable to check the logic of the instructions. These error messages so produced by the compiler are known as *diagnostics*. The process of correcting the diagnostics is known as *debugging*.

Diagnostic errors produced are limited to those in the construction of program statements. For instance in a COBOL program the computer will not accept 'deduct' in lieu of 'subtract'. Any errors of logic in the construction of the program will not be revealed by this process. It is possible for a program to be correct in terms of its language and still not be able to process data correctly. Test data must therefore be prepared in order to test the logic of any program. If satisfactory results are not obtained the instructions must be analysed and suitable amendments made to the source program. The object program must

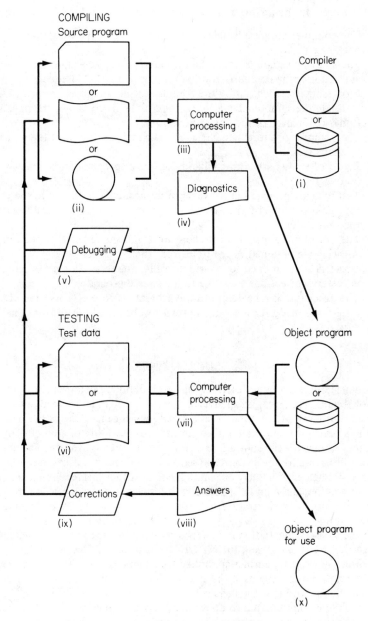

COMPILING
Source program

Compiler

or

Computer
processing
(iii)

(i)

Diagnostics
(iv)

(ii)

Debugging
(v)

TESTING
Test data

Object program

or

Computer
processing
(vii)

or

(vi)

Corrections
(ix)

Answers
(viii)

Object program
for use

(x)

Figure 5.1   Compiling and testing programs

45

then be discarded and the whole process of compilation repeated using the revised source program.

The process for compiling and testing programs is as follows (see figure 5.1).

(i) The compiler program is fed into the core store of the computer from a backing store.

(ii) The source program is transferred to punch card, paper tape or magnetic tape and fed into the computer for due processing by the compiler.

(iii) Processing is carried out by the computer using the compiler program to translate the symbolic language of the source program into the machine (binary) language of the object program.

(iv) The compiling routine will check errors of syntax and print them out (diagnostics).

(v) Errors of syntax are corrected and the source program altered accordingly (debugging).

(vi) Test data are prepared in the form of input media and fed into the computer for due processing by the newly created object program to test the logic of the operations.

(vii) The object program is fed into core store and the test data processed.

(viii) Answers are obtained as print-out from the computer.

(ix) Answers are examined for spurious results due to an illogical program; all necessary corrections are made to the source program.

(x) The procedure is repeated until the program is correct; corrected object program is ready for use when correct results are obtained from the test data.

SUBROUTINES

A sequence of instructions describing any well-defined operation and forming part of the main program is called a subroutine.[1] Programs are made up of defined routines that are suitably linked in a logical order. One of the objectives in programming is therefore directed towards the assembly of a number of suitable routines for the solution of a problem. The reason for using a subroutine is to avoid having to repeat the same sequence in different places in the master routine. A single routine may be simultaneously a subroutine with respect to another routine, and a master routine with respect to a third. The control of a single subroutine may be transferred from more than one place in the same master routine.

In the COBOL illustration certain routines can be seen under the headings of AREA-CALCULATION and PRINT-OUT. The program flowchart (figure 4.6) also shows how these routines are linked (see connectors). A decision is usually required at the beginning or at the end to control entry and re-entry. In this instance the control is at the beginning of the routine (AREA-CALCULATION) and tells the computer what to do when all the cards have been read. The *looping* so formed by the last instruction of this operation causes the sequence to be repeated until the condition is reached which requires a decision. On completing the last card the processing is switched to PRINT-OUT.

Subroutines may take different forms, as follows.[9]

46

(a) *Direct Insert or In-line Subroutine*   Certain types of subroutines are inserted directly into the linear operational sequence of instructions rather than by a jump. In such instances they must be recopied at each point that they are needed.

(b) *Unlinked or Closed Subroutine*   Routines that are not stored in the main path of the program are entered by a jump operation. This is an operation that alters the logical linear sequence of the main program. While the computer is performing the subroutine the main program is interrupted and provision is made to resume when the operation is complete. Instructions relating to the entry and re-entry into the linear operations constitute the linkage.

(c) *Scientific Subroutines*   There are several scientific subroutines that perform standard mathematical operations. Extensive libraries of such subroutines are delivered with every computer system for use by computer clients.

It is possible to write a program in a high-level language that contains a subroutine in another programming language. The computer is instructed to find the required subroutine and assemble it within the program, for the solution of the problem.

MACRO INSTRUCTIONS[10]

Instructions may be given to the computer that will generate several machine instructions calling for special routines or library routines to perform wanted functions like opening, seeking and closing files. The facility also permits the break up of a large program to allow portions to be tested separately. This is an extensive program analysis that is an aid to debugging.

A *macro code* is a coding system that assembles groups of computer instructions into single code words. It uses the one-to-many concept and generates several machine language instructions from one macro statement in a source program. Many tedious and meticulous chores can be lessened by the use of macro codes.

With the aid of macro instructions the computer becomes its own programmer or translator. For example, the production of a bill of quantities requires a number of different programs that have to be introduced separately. If backing stores containing the various programs are set up on line the human element of having to introduce the programs separately after successive completion may be eliminated with the aid of macro instructions. The data are fed in during program number 1 and the rest is a machine operation for the successive selection and operation of the various programs.

REFERENCES

1. J. Maynard, *Computer Programming Made Simple* (W. H. Allen, London, 1972).
2. T. F. Fry, *Computer Appreciation* (Butterworth, London, 1970).
3. S. H. Hollingdale and G. C. Tootill, *Electronic Computers* (Penguin, Harmondsworth, 1966).

4. A. Learner and A. J. Powell, *An Introduction to ALGOL 68 through Problems* (Macmillan, London and Basingstoke, 1974).
5. C. J. Sass, *FORTRAN IV Programming and Applications* (Holden-Day, San Francisco, 1974).
6. J. Watters, *COBOL Programming* (Heinemann, London, 1972).
7. W. G. Dodds, *Basic COBOL* (IPC Electrical–Electronic, London, 1970).
8. M. Bohl, *Information Processing* (Science Research Associates, Henley-on-Thames, 1971).
9. C. J. Sippl, *Computer Dictionary* (Sams, Indianapolis, 1966).
10. J. Maynard, *Computer Programming Management* (Butterworth, London, 1972).

6 INFORMATION RESOURCES

If computers are to be of any use to quantity surveyors and the construction industry as a whole, programs must be written with the needs of the user in mind. The information needs of the industry must therefore be examined to establish the areas where computers would be most useful. In order to assess this usefulness the complexity of communications within the industry must be untangled. By this means it is possible to define the problems that might be satisfactorily solved by computer usage.

THE BUILDING PROCESS

The building process is a complex operation involving individuals or groups of individuals, each performing specialised procedures. The building process as a whole comprises a number of sectors that range from (a) a design team that looks after the client's interests and requirements, to (b) the manufacturers and suppliers of materials and components and (c) the contractors and sub-contractors who execute the work. The design team comprises the following members.

(i)   The architect who is responsible for ensuring that the efforts of all concerned in design and construction are brought together at the right time and in the right sequence. He is also responsible for ensuring that the work performed in design and construction is in accordance with the specified requirements.
(ii) The quantity surveyor who is responsible for giving advice to designers on matters relating to costs and contracts for building and civil engineering projects.
(iii) Consulting engineers who advise on many aspects of engineering work; such consultants include structural engineers, services engineers and acoustics consultants. The structural engineer is concerned with prescribing the conditions for structural efficiency, stability and structural form of particular projects. Services engineers are concerned with the environmental control of a project such as heating, lighting and the provision of utilities such as lifts. Acoustics consultants advise the architect in acoustic measurement, calculation and prediction based on their knowledge of sound and noise control design techniques.

The building process begins when the client seeks advice from the designers, and it continues through a sequence of operations to the completion of the project. It may be considered as a linear sequence of procedures from its inception to completion as follows.

(a) Inception
(b) Feasibility
(c) Outline proposals

(d) Scheme design
(e) Detail design
(f) Production information
(g) Tender action
(h) Project planning
(i) Operations on site
(j) Completion
(k) Feedback

Inception

At the inception of the project the requirements of the client must be defined.
In the early stages his ideas will be vague. The client will usually call upon the
architect to prepare a brief giving guidelines for the preparation of a general
outline of requirements and a plan for future action. In so doing his ideas will
become more crystallised. The brief will contain the client's requirements by
stating the performance expected of the project with any restraints on the
freedom of design. It will also contain particulars important to the client, such
as cost limits and timing.

Feasibility

After the client's requirements have been established a feasibility study is carried
out on such aspects as finance, practical situation and user requirements. The
architect and quantity surveyor will study the particulars contained in the brief,
make an appraisal and provide recommendations, so that the client can determine
the form the project is to take in the event of it being a feasible proposition. The
architect, engineers and quantity surveyor will carry out the necessary studies of
user requirements, planning, design work and cost evaluation.

Outline Proposals

If and when a decision is taken to proceed with the project and a cost limit is
established the architect will develop the brief. At this stage outline proposals
and a report are prepared concerning the layout, design and construction of the
project in order to obtain the approval of the client. Cost studies of alternatives
may also be made by the quantity surveyor taking into account shape, height,
storey height and span. An outline cost plan may be prepared using a brief cost
analysis to confirm the cost limit.

Scheme Design

This involves the preparation of sketch plans and preliminary estimates of costs.
The form of construction and the plan must be established with the position of
key rooms, stairs and accesses, together with the proposed services. An outline
specification and a full cost plan with further cost studies may be prepared.
Preliminary designs must be obtained from specialists. At the completion of the

scheme design the client should have

- (i)   a visual realisation of the building
- (ii)   specialists' proposals
- (iii) an outline specification
- (iv) an elemental breakdown of costs.

Detail Design

The architect, in collaboration with engineers and other specialists, and the contractor and subcontractors if appointed, will prepare a full design of the building. Complete cost checking of design will also be made by the quantity surveyor. This enables a final decision to be obtained on design, specification, construction and cost. Any changes made after this stage has been reached will result in abortive work.

Production Information

After the detailed design decisions have been made the final production information is prepared. This consists of drawings, schedules and specifications by the architect and his consultants and the preparation of bills of quantities and tender documents by the quantity surveyor.

Tender Action

Estimates are prepared by contractors and tenders are submitted. The action taken should be in accordance with the 'Code of Procedure for Selective Tendering' published by the National Joint Consultative Committee of Architects, Quantity Surveyors and Builders.

Project Planning

The contractor and subcontractors will prepare programs of work for the site operations.

Operations on Site

Building operations are carried out by the contractor and subcontractors for the construction of the project under the supervision of the architect. Work is valued by the quantity surveyor for the purposes of interim payments and reports are prepared on the financial effects of variations.

Completion

On practical completion the building is handed over for occupation to the client or his tenants. During the defects-liability period, usually lasting six months after completion of the work, the contractor must remedy all defects that appear. When all work is satisfactorily completed a final certificate is issued. This is based on a bill of variations prepared by the quantity surveyor.

Feedback

An analysis should be made of job records and a study made of the building in use. This will provide useful information for the architect, quantity surveyor and engineers on future projects. An analysis of management, construction and the performance on the project will also provide useful information for the contractor and subcontractors on future projects.

Every building project is performed by the execution of the above procedures. These can be grouped into functions marking the achievement of an identifiable objective, namely the functions of design, design realisation, construction and management. Hence each of the three functional groups is composed of sets of procedures that do not depend on the structure of the industry (figure 6.1).

The design function involves the procedures that cover the definition of the client's needs, feasibility, outline proposals and scheme design. The operations for this particular function will produce (a) a brief, (b) a sketch design and (c) a cost plan. Participants will be the client, architect, quantity surveyor and engineers.

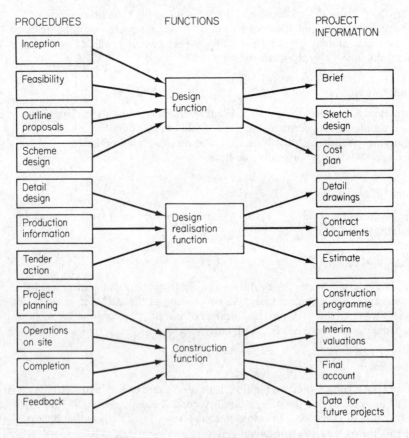

Figure 6.1   The building process

52

The main work undertaken by the design team in performing the function of design realisation is that of production information. Design realisation starts once the design function has resulted in the choice of structural techniques and commodities. Production information consists of the preparation of production drawings, specifications and schedules and bills of quantities. This function involves detail and design and production information procedures and is concerned with the final selection of products, the preparation of production information and estimating. The outcome of design realisation is a description of the project and an estimate of cost. This information should be arranged in such a way that the contractor and subcontractors can use it directly without further measurement or reallocation. Any rearrangement should be a simple mechanical process. Participants in this function are the architect, quantity surveyor and the engineers.

The function of construction deals with the contractor's activities in the performance of his obligations under the contract. This concerns the procedures involving project planning and operations on site. The result of project planning is a construction program that is used as a guide to the performance of the building work. Valuations will be made by the quantity surveyor as the work proceeds and also at completion.

The management function is concerned with the planning and control of the building process. This function embraces each of the functions of design, design realisation and construction and also covers the management of organisations of whatever kind.

The architect is the leader of the design team that has divergent interests and technical skills. The divergent nature of the various members of the team has led to a certain fragmentation with poor communications that tend to hinder the vital interaction between the architect and his consultants. Good communications and a good flow of information between the participants in the building process are vital.

COMMUNICATIONS AND FLOW OF INFORMATION

Communication is the means of transmitting ideas and information. It is concerned with the information needed, the information that results and the functions from which and to which the information passes. Success or failure in communications is the sole responsibility of the initiator and is dependent upon the ability of individuals with differing technical skills and sectional interests to pass on ideas.

Communication is the binding force of the building process, the lack of which may result in a duplication of effort – the same information might be processed repeatedly by different people. Information produced by any member of the building team is likely to be produced independently by others. This will happen particularly where members of the team are unaware that the data exist or where the data are inaccessible or not in the right form for their requirements – this is wasteful in both effort and expense.

The communication problem is the transfer of ideas from the designer through the various operations to the contractor or subcontractors without any change of meaning. Each participant in the process deals with details in a different mode and the vehicles used to communicate information vary with the operations performed. The weakness of communication lies with the differing modes of transfer. Information is made available in a number of different ways. These may be verbal, descriptive, graphical, models or codes.

Verbal transmission of data is the most common form of communication. This would generally only relate to descriptions and demonstrations appropriate to operatives.

Descriptive communication takes the form of written information such as schedules of requirements, specifications, bills of quantities, letters and contracts. This form generally supplements the graphical form of communication.

The graphical transmission of information is one of the main vehicles of communication from the architect and takes the form of drawings using orthographic or pictorial representation. Another popular method of communication, particularly for the benefit of lay people, is the use of models. This method consists of a mock-up or the construction of a model of the project. Samples of materials may also be submitted to illustrate quality. For practical purposes templates, jigs or storey rods may be used.

Coded information on punched card, paper tape, or magnetic tape is being used in connection with communications through computer usage. The computer's ability to sort, process and store information is probably more significant than the speed at which it makes calculations. In this context the computer should be treated as a means of communication.

The particular vehicle of communication depends on the nature of the data to be recorded and transmitted. The flow of information is related to the various functions performed during the building process, rather than by whom they are performed. The architect is concerned with information on the planning of the project and details are passed on in the form of drawings. Quantity surveyors are concerned with providing information relating to contract documentation that supplements the architect's information; this will be written. The contractor is concerned with using all this information for the purpose of translating the ideas passed on by the design team.

Information should be passed on in a manner that is most useful to the recipient and should provide quick and easy referencing. The architect sometimes produces drawings that are not organised with any particular pattern in mind and do not bear any relation to the sequence of operations on site; their form may preclude any re-use on future projects. Only bills of quantities can be said to have a format that will help the contractor find information without any lengthy search. Certain defects, however, do exist in respect of the format of traditional bills of quantities.

Information produced by some participants as output may form the basic data input to other participants. Basic data have little value by themselves and the more they are used in the decision-making processes the more they increase in value. Drawings are produced that are output to the architect and basic data input to the quantity surveyor and the contractor. The quantity surveyor produces information for a number of purposes, such as tendering and the post-contract needs of the contractor, the architect and himself. The contractor uses the drawings for project planning and site operations.

It is important that a satisfactory interrelationship exist between each participant for the supply of information that facilitates easy processing. For information to be of any use to the various members of the construction industry it must be properly coordinated. This means that information should have a corresponding relationship and should be presented in consistent terms having some connotation within the industry. Some form of data coordination must be maintained in the building process to ensure good communications.

DATA COORDINATION

This chapter is not concerned with illustrating any system of data coordination but rather with the examination of the factors involved and their significance in computer applications. Data coordination is a method of rationalising the means and contents of communications. It is the organisation and control of factual information that establishes a relationship of data. This is also a prerequisite to an effective use of the computer as a vehicle of communication. Beneficial results should be obtained by using such a system, since it will lead to a better use of information, or more economical working, or both. The advantages of data coordination may be summarised as follows.[1]

(a) It provides a common language for use between participants, which reduces the risk of misunderstandings.
(b) A more effective coordination of effort is possible between participants.
(c) It provides a design check ensuring that no requirement has been missed.
(d) Past and present projects may be more accurately compared.
(e) Libraries of standard or type details and clauses may be built up to help ensure that work is not duplicated.
(f) It assists programming, design and document production.
(g) It permits cross-referencing between all documents.
(h) It facilitates the use of computers and makes it possible to sort information in different ways to meet different needs. Data coordination and computer applications are synonymous inasmuch as the ingredients for an effective system for the organisation and control of information are those necessary for a good computer-based system of information control.

The criteria applicable to schemes for data coordination and coding were set out in a report by the Building Research Establishment[2] as follows.

(i) Classify information
(ii) Identify and describe resources
(iii) Describe projects
(iv) Foster the development of procedures
(v) Support information flow

Classify Information   This is necessary to enable information to be filed, sorted and retrieved in ways useful to the industry. The classification categories should be compatible with the user's needs.

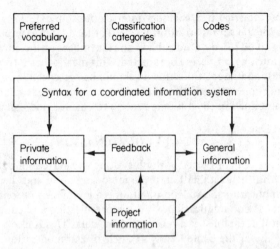

Figure 6.2   Framework for a coordinated information system

Identify and Describe Resources   This concerns the transactions between participants and involves the description of resources about commodities, plant and equipment.

Describe Projects   An adequate description of a project is the most important feature of a coordinated information system. Projects are described by using one or a number of the vehicles of communication that are usually in a graphical or descriptive form. Graphical descriptions usually consist of drawings; written descriptions include schedules, specifications and bills of quantities.

Development of Procedures   This involves communications and the need for improved arrangements in order to provide an effective system of data coordination procedures to make data more accessible. This means standardising the presentation of data by agreed methods of working. The aim is to provide categories for the retrieval of data and for the arrangement of documents in ways that will suit the needs of the users and provide a means of coordinating the conventions evolved for and by each participant in the building process.

Support of Information Flow   This role is fulfilled if the foregoing rules are substantially discharged. However, circumstances must be created that lead to a good information flow with a proper coordination. Information relating to the building process may be classified into three main categories. These are considered to be private, general and project information.

Private information consists of office standard details and cost and output records, which are usually confidential to the participants involved. Clients, designers, quantity surveyors and contractors will each have a store of private information.

General information consists of that type of information that is available to everyone and applicable to the project. It consists of codes of practice, building regulations, manufacturers' catalogues, standard methods of measurement and the like. Much of this information is scattered and difficult to locate and assemble; it is best held on a commodity file for general use.

Project information is that information created for a particular project. It is created afresh by the participants on each project. It comprises such details as the client's brief, production drawings, specifications, schedules, bills of quantities and the like. All the information that is particular to a project and available to those engaged on it is project information.

The flow of information is interactive between these three categories. Private information is used with general information to produce project information. Project information in its turn may form a feedback to private information.

Framework for Coordinated Information Systems

The construction industry is large and diverse and changes often occur in the building process. This makes it impossible to develop any comprehensive system of data coordination to serve all the functions discharged. Data coordination should therefore consist of a number of coordinated information systems serving specialised and limited objectives. These are best related to the procedures outlined for the building process and consist of the production of information to serve a function of the building process.

The framework for any coordinated information system has a syntax or a set of rules, which consists of three basic elements

(1) preferred vocabulary
(2) classified categories
(3) codes.

These help to coordinate and integrate the three categories of information (that is private, general and project) from which each participant will draw for the purposes of performing the procedures necessary to fulfil a particular function (figure 6.2).

A preferred vocabulary means a standardisation of expression and is an essential part of the basic syntax for both project and general information. A major feature is a vocabulary of words recommended for standard use. This arises out of the need to classify information that will only be meaningful if the words and phrases used have exact and unambiguous meanings. There are many sources of reference, such as the British Standard specifications and Codes of Practice, standard forms of contract and standard methods of measurement from which a vocabulary can be drawn.

A report on data coordination by the Department of the Environment[3] recommends a basic syntax for information systems. The first major element is considered to be the preparation of a glossary of terms for use in communication. This was recommended as a means of coordination since any schedule of words needs to be accompanied by definitions to assign a precise and unique meaning to each word or phrase.

Classification of information is a second major element of the basic syntax and consists of classification categories of groups of information. A well-known classification is the CI/SfB system, which is the British version of the

International Council for Building Research Studies and Documentation (C.I.B.). This system has been recommended by the Royal Institute of British Architects for use as a means of arranging project information. It has also been used in the structuring and coding of the National Building Specification. The construction industry uses many classification categories ranging from filing systems to the various arrangements of formats for bills of quantities.

A third major element is a system of codes so that information can be concisely identified with precision. Coding is important from the point of view of commodity identification. It also becomes necessary with the growing use of computers.

CODING

Natural languages contain ambiguities and a choice of modes of expressions that make their use impractical for communications with the computer. A more comprehensible form of communication between man and computer involves restricted vocabularies comprising preferred lists of symbols with no ambiguities of meaning – these are called codes. A code is a set of symbols that may be grouped or ungrouped to represent information or data and to facilitate their storage and transmission or manipulation. Coding in this context must be distinguished from the computer languages used to specify computer programs.

The object of using codes is to

(a) compress information
(b) increase accuracy of transmission and interpretation
(c) enable a variety of types of manipulation to be performed.

The process of writing coded data from its original form of information is known as encoding; the reverse process is known as decoding. Both encoding and decoding require the application of a set of rules known as *code keys* to enable the processes to be performed.

For a code to be effective it must have a format that contains recognisable *fields*. These are groups of symbols each of which has some significance. For instance English words may be considered to be fields of various lengths. It is difficult to use natural language words since they are of variable lengths, also the code would not be international and difficulty might be experienced in selecting words that have the same meaning in different branches of the construction industry. However, interpretation may be made without reference to a code key although reference may have to be made when selecting preferred words. A complete code structure contains fields of fixed lengths but these may vary according to the significance of the field.

Types of Code

The following types of code are used for manual or computer purposes.

 (i) Mnemonic codes
 (ii) Numeric and alphabetic codes
(iii) Binary codes

Mnemonic Codes   A mnemonic code is an instruction code that uses conventional abbreviations. ('Mnemonic' is from the Greek word for memory.) This type of code has advantages when coding and decoding and is easily interpreted without reference to code keys. The mnemonics of natural language words have the advantage of compactness and are therefore more economical for the transmission and storage of information. This type of code is used for both manual and computerised processes. The taker-off uses it when measuring and the programmer may use it when writing computer programs.

Examples of mnemonic coding are as follows.

(1) when used by the taker-off, basement becomes 'bast', concrete becomes 'conc', deposit becomes 'dep'.
(2) when used by the programmer, multiply becomes 'mult', subtract becomes 'sub', transfer becomes 'tran'.

There are no standard mnemonic codes common to all users since each usually develops his own. This can lead to confusion and ambiguities when two or more words are used giving the same mnemonics. More complex rules may be used to avoid this but any advantages in ease of formation and interpretation will be lost.

Numeric and Alphabetic Codes   This type of coding consists of symbols that may be numeric, using numbers or digits as the characters of the code, or alphabetic (alpha), using only letters of the alphabet and special characters of full stop, comma, asterisk and others as the characters of the code. These codes may be used for communicating information between manufacturers, designers and constructors. A code structure using these characters may be numeric, alphabetic or alphameric; an example of this last is CI/SfB classification. Numeric and alphabetic characters are also used for computer programs.

The advantages of using only numeric characters are as follows.

(1) The keyboard is likely to remain the chief method of preparing input data for quantity surveyors for some time and less training is required for typists; typing speeds may also be higher.
(2) Numerical coded information may be considered to be easily placed in order.
(3) Calculating machines and small computers usually deal with numerical information more naturally than alphabetic characters.
(4) Cheaper machines may be produced that deal with a limited set such as numerical characters.

Certain disadvantages of using numeric characters may be considered as advantages for using alphabetic characters, as follows.

(1) Humans, and to a lesser extent computers, are able to distinguish non-numeric symbols more easily than particular sequences of numbers or spaces contained in a numeric code.
(2) A combination of letters and numerics enables different fields of a code to be distinguished without the use of extra tags to each field; fields using only letters or numbers must be of different lengths or carry a distinguishing tag.
(3) If a field is easily distinguished it eliminates the need for fixed positions relative to one another in documents.

(4) Sets of characters using a mixture of letters and numbers are more easily memorised.

Binary Codes   Binary coding is suitable for information requiring a yes/no answer to a question. This coding system is useful for storing information for computer use. The bistable nature of the machine is utilised and a binary code is used for the storage of information.

Code Fields

The fields of a code used for processing information by computers may have (i) a fixed format for the code field and (ii) a fixed position of the code field.

A fixed format for the code field means that it is of a particular fixed length. This is denoted by the number of digits. The order in which the letters and figures or other symbols appear within the field may be irrelevant. Thus a field may have a length of four symbols, for example F463, AJ93, 7H43 and 5HD8; although these have the same field length they do not have the same format. Coded information with a fixed format is easy to follow and any missing or added characters within the fixed field are obvious. One major disadvantage of having a fixed format is the need for excessive spare space within the fields to allow for future development.

A fixed position of the code field means that each field has a fixed position in the code structure. This is important for the coding of standard descriptions. For instance, each clause of a standard description constitutes a code field and the order in which it appears depends on the level of the clause in the description.

Example   The code structure for a standard description could be as follows, where 'A' denotes a capital letter or alphabetic code and 'N' denotes a numeric digit.

Level 1 clauses have a code field with a format – A
Level 2 clauses have a code field with a format – AA
Level 3 clauses have a code field with a format – NN
Level 4 clauses have a code field with a format – AANN
Level 5 clauses have a code field with a format – NAAN

The code would then appear in the following order

A AA NN AANN NAAN

Using actual symbols the code might appear

F BE 45 FE26 3ST5

The advantage of using such a code structure is that simpler routines may be developed for producing, checking and interpreting coded information. The position of the code fields may also be easily distinguished and this is a safety factor where each type of field has a different format. The disadvantages of having fixed order and number fields may mean an excessive redundancy where some fields rarely carry information. Errors of misplaced digits may have repercusssions throughout the code.

COORDINATED INFORMATION SYSTEMS

Coordinated information systems have been developed for use in the various stages of design, documentation and construction.[4] Systems are usually based on the SfB coding and allow coordination with related documents and the retrieval of data. They are considered to be complete identification systems for the building process. Every item is uniquely identified and carries the same references throughout the whole of the project documents. The key to data coordination and information retrieval is the code structure. This consists of two main fields, which are classified as (a) the general code (b) a number of specific codes.

The *general code* classifies all relevant labour and material processes in terms of form or structure and is meant to be an industry code that is general to all projects. This code field is based on the SfB facets and classifies the different processes according to the nature and location of the item.[1] It is divided into a number of facets as follows.[5]

1. Facet   This signifies a location and classifies the building elements to which the item belongs. For instance

  (2–) SUPERSTRUCTURE
  (20) SITE SUPERSTRUCTURE
  (21) External walls
  (22) Internal walls
  (23) Floors

2. Facet   This signifies a form and classifies the construction of the item using symbols A to Z. For instance

  E – Concrete constructions
  F – Brick constructions
  G – Structural unit constructions

3. Facet   This provides a further classification to indicate the resources (labour and materials) of the item. For instance

  e – Natural stone
  f – Formed (precast) concrete, etc.
  g – Clay in general

This is further amplified as follows

  g   – Clay in general
  g1 – Adobe, cobe, pise (rammed earth, moler earth, etc.)
  g2 – Fired clay
  g3 – Faience (glazed fireclay, ceramic, terracotta, etc.)

4. Facet   This symbolises the particular process or element within its class and is known as the identification part of the general code, thus providing a specific identity. The allocation of '4 Facet' numbers is based on numeric references of well-defined items. For instance, assume

  g2.1010 – Clay common bricks to BS 3921 Part 2 special quality g2.1020 — Clay hollow blocks to BS 3921 Part 2 Chapter 1 keyed for plaster

g2.1030 — Multicoloured red rustic facing brick

Example   In the notation (21)Fg2.1010

(21) — signifies a location (external walls)

F — signifies a construction (brick construction)

g2 — signifies a material (fired clay)

1010 — signifies the specific identity of the material (clay common bricks to BS 3921 Part 2 special quality)

The code thus signifies work in EXTERNAL WALLS OF BRICK CONSTRUCTION WITH CLAY COMMON BRICKS TO BS 3921 PART 2 SPECIAL QUALITY.

The *specific codes* attempt to locate and assemble the items into groupings for practical execution of the project. The general code merely identifies the work, not its location. A further set of codes is necessary to identify a particular activity in relation to the project in question. The code fields used are as follows.

Job Code

This identifies the contract, for example 'CO64'.

Block Code

This identifies a particular block; thus for a project for the construction of a university the following might apply.

A1   Administration block

B1   Faculty of Arts

B2   Faculty of Law and Social Sciences

B3   Faculty of Applied Science

Thus if the item is part of a block housing the Faculty of Arts the reference for the block code is 'B1'.

Storey Code

This identifies the item in relation to the storey or floor level of the block in question. Thus for a five-storey block the references might possibly be

00   General

01   Ground floor

02   First floor

03   Second floor

04   Third floor

05   Fourth floor

If the item is on the third floor, the reference is '04'.

Section Code

This identifies the item in relation to any subdivisions of any storey. For instance the following divisions might apply.

A0   General
A1   Administration section
A2   Research section

If the item is applicable to the whole floor, the reference is 'A0'.

Room Code

This identifies the individual rooms on each floor. For instance

300   General
301   Room 1 on the third floor
302   Room 2 on the third floor
303   Room 3 on the third floor

and so forth.

Feature Code

This reference is used to identify certain constructional parts such as walls, stanchions or beams that need to be identified on the drawings to simplify construction organisation. The feature is associated with a set of elements. For the purpose of illustration, assume the item to be infill panels reference 'IP4'.

Trade Code

This identifies the skill or trade required to execute the item. For the purpose of illustration, assume this to be 'GO1'.

Activity Code

This identifies the activity to which the item belongs. These codes usually identify activites in PERT/CPM networks (chapter 10). For the purpose of illustration assume the reference to be '036'.

From the foregoing references the specific code would appear thus

CO64 B1 04 A0 303 IP4 GO1 036

The complete code would appear thus

(21)Fg2.1010 CO64 B1 04 A0 303 IP4 GO1 036

The code then signifies an item in activity number 36 involving work in external walls of brick construction forming an infill panel (IP4) of clay common bricks to BS 3921 Part 2 special quality. The work is situated in room number 303 on the third floor of the Faculty of Arts block for contract number CO64.

The type of code for computer use will depend on the nature of the documents produced. The following categories of document are used.

(i) Input Documents   These are documents that are intermediate between the user and computer and are used as data input. Codes used for input documents should be short, for economy of encoding procedures and computer input time. They should also be appropriate to the skills of the persons involved. There should be a minimum of look-up time when referring to the code keys. Code symbols should be appropriate to the various input devices.

(ii) On-file Documents   These consist of coded information stored on computer media that are intermediate between input and ouput. They comprise such devices as magnetic tape, discs and drums. The codes should be short for economy of storage and computer cycle time, and should be suited to the types of file processing such as retrieval, sorting and output.

(iii) Output Documents   These are products of the computer. Codes should be easy to decode and appropriate to the skills of the user.

Computerised Information Systems

The use of an electronic digital computer may be considered to be a key factor in implementing a coordinated information system. Its advent has made possible a higher degree of storage, processing and retrieval. There are certain limitations since such systems are geared to the operating power of the central processor. However, with the rapid progress of computer technology the scope for the development of a coordinated information system is not limited.

The report *A Study of Coding and Data Coordination for the Construction Industry*[2] discusses the type of information system that might be incorporated into a future information system of the construction industry. Features of such a system embrace the compilation of file documents with coordinated sources of information. The major file documents comprise (i) project file, (ii) contractor's general file, (iii) contractor's private reference file, (iv) central commodity file, (v) designer's reference file, (vi) national standards, regulations and specifications and (vii) technical information file. These may conveniently be stored on file-holding devices such as magnetic disc, exchangeable disc, magnetic card file and magnetic tape. The key factors that control the speed and development of the type of computerised system suggested are (1) the costs and capabilities of the storage of information on the computer and (2) the costs of communication between the files of data or between users.

REFERENCES

1. A. Ray-Jones and W. McCann, *CI/SfB Project Manual: Organising Building Information* (Architectural Press, London, 1971).
2. ——*A Study of Coding and Data Co-ordination for the Construction Industry* (Building Research Establishment, H.M.S.O., 1969).
3. ——*An Information System for the Construction Industry: Final Report of the Working Party on Data Co-ordination* (Directorate General of Development [Housing and Construction] Department of the Environment, H.M.S.O., 1971).
4. ——General Brochure, Construction Control Systems Ltd.
5. ——*Introduction to the CBC System* (Co-ordinated Building Communication, CBC Publications, 1969).

7 STANDARDISATION OF INFORMATION

Like many other industries, it has been predicted that computer applications will have a profound effect on the way the construction industry operates. The development of computer techniques has provided a strong impetus towards the standardisation of contract documents and project information.

Project information systems involve the assembly, cross-referencing and preparation of drawings, bills of quantities and other documents. The syntax for quantity surveyors' coordinated information systems entails a standardised structure for (a) methods of measurement, (b) descriptions and (c) the coordination of manual and computerised methods of presenting information.

METHODS OF MEASUREMENT

The main types in use are as follows.

   (i) Standard Method of Measurement of Building Works[1]
  (ii) Code for the Measurement of Building Works in Small Dwellings[2]
 (iii) Standard Method of Measurement of Civil Engineering Quantities[3]

Standard Method of Measurement of Building Works (S.M.M.)

This is the main source of reference used by the quantity surveyor, which involves standardisation. It is a document that provides a uniform basis for the measurement of building works. Prior to 1922 there was a lack of uniformity in methods of measurement and a diversity of practice that varied with local custom. This situation formed a breeding ground for uncertainty and the true meaning of items in bills of quantities were left in doubt. In consequence, frequent demands were made on the services of the Surveyors Institute (now the Royal Institution of Chartered Surveyors) and the Quantity Surveyors Association for clarification. As early as 1909 these bodies were aware of the necessity for securing some form of standardisation in measurement and they published pamphlets and circulars giving opinions on the correct methods of measurement concerning disputes that arose. Cooperation between the two bodies became a fact in June 1912 when a Joint Committee was set up to draft a comprehensive set of rules. Representatives of the building trades were added to this committee in 1918. The committee then comprised six surveyors (nominated by the Surveyors Institution and the Quantity Surveyors Association) and four contractors (nominated by the National Federation of Building Trades Employers and the Institute of Building). The two surveyors' professional bodies amalgamated in 1920 and subsequently published the first edition of the S.M.M. in 1922.

Code for the Measurement of Building Works in Small Dwellings

Early in 1944 the R.I.C.S. and N.F.B.T.E. decided to explore the possibility of evolving some simple system of measurement for the superstructures of small dwelling houses. The first edition was published in 1945 and is designed to meet the requirements of builders experienced in pricing such works. Composite descriptions are used, thus reducing the number of items in the bills of quantities. Direct reference is made to the S.M.M. and all items in that document are deemed to be applicable unless otherwise indicated in the code.

Standard Method of Measurement of Civil Engineering Quantities

The general principles used for the measurement of civil engineering work cater for the fact that civil engineering bills of quantities are read in conjunction with the drawings. The descriptions attached to the items in the bills of quantities should only be in sufficient detail to ensure the identification of the work covered by the respective items shown on the drawings and described in the specification. Descriptions in the bills of quantities should not repeat unnecessarily the descriptive matter contained in the other documents. The drawings, specification and bills of quantities are intended to be read together.

STANDARD DESCRIPTIONS

A development that has been adopted by many quantity surveyors is the use of standard descriptions for bills of quantities. This is achieved by arranging the constituent terms of any 'item' description into a graded structure. The following descriptions will illustrate the principles involved when their graded structures are analysed.

Description 1

'PLAIN INSITU CONCRETE (1:2:4/40 MM AGGREGATE) FOUNDATIONS NOT EXCEEDING 150 MM THICK IN TRENCHES'

Description 2

'BRICKWORK IN COMMON BRICKS IN CEMENT MORTAR (1:3) REDUCED TO ONE BRICK THICK IN PROJECTIONS'

The graded structure of any item description consists of a number of 'levels' each of which contains alternative words or phrases. The description 'levels' are referenced as follows.

Level 1 — The main section heading (usually follows the S.M.M.)
Level 2 — Subsidiary division within the main heading
Level 3 — Specification of materials and workmanship
Level 4 — Identification of work involved
Level 5 — Variables such as size and colour

Each level of a description contains a number of phrases that are interchangeable within the same level.

Level 1 A description is first classified according to the main section to which it belongs. Such classifications may be as follows: (i) Demolitions and Alterations, (ii) Excavation and Earthworks, (iii) Piling, (iv) Concrete Work, (v) Brickwork and Blockwork, (vi) Underpinning, (vii) Rubble Walling, (viii) Masonry, (ix) Asphalt Work, (x) Roofing, (xi) Carpentry, (xii) Joinery, (xiii) Structural Steelwork, (xiv) Metalwork, (xv) Plumbing and Engineering Installations, (xvi) Electrical Installations, (xvii) Plasterwork, (xviii) Glazing, (xix) Painting and Decorating, (xx) Drainage, (xxi) Fencing. *Note* Phrases at this level are not interchangeable. From the foregoing the following will apply.

Description 1 – CONCRETE WORK
Description 2 – BRICKWORK AND BLOCKWORK

Level 2 This second level of classification is concerned with the subsection of the main section to which the description belongs. The subsections of 'Concrete Work' may be classified as follows: (i) Plain *in situ* concrete, (ii) Reinforced *in situ* concrete, (iii) Labours on concrete of any description, (iv) Reinforcement, (v) Formwork to plain *in situ* concrete, (vi) Formwork to reinforced *in situ* concrete, (vii) Precast concrete, (viii) Sundries.

The subsections of 'Brickwork and Blockwork' may be classified as follows: (i) Brickwork, (ii) Brick facework, (iii) Brickwork built entirely of facing bricks, (iv) Blockwork, (v) Damp-proof courses, (vi) Sundries.

From the foregoing the following will apply.

Description 1 – PLAIN INSITU CONCRETE
Description 2 – BRICKWORK

Level 3 The third level covers the variations of materials and workmanship. For instance, the materials and workmanship clauses as follows.

(a) 'Concrete Work' covers such factors as
    (i) type of concrete such as normal, sulphate-resisting, water-repellent, lightweight, or insulating
    (ii) concrete mix, such as 1:12, 1:3:6, 1:2:4
    (iii) size of aggregate, such as 40 mm, 20 mm, 10 mm.
(b) 'Brickwork' covers such factors as
    (i) type of bricks, such as common bricks, calcium silicate bricks, engineering bricks
    (ii) type of mortar, such as cement lime, cement
    (iii) mortar mix, such as 1:2:9, 1:1:6, 1:3.

From the foregoing the following will apply

Description 1 – NORMAL CONCRETE MIX 1:2:4/40 MM AGGREGATE
Description 2 – COMMON BRICKS IN CEMENT MORTAR (1:3)

Level 4 This level is concerned with the remaining portion of a description that identifies the work involved.

67

(a) For 'Concrete Work' such items as
   (i) mass filling
   (ii) foundations
   (iii) ground beams
   (iv) machine bases
   (v) beds.
(b) For 'Brickwork' such items as
   (i) reduced to one brick thick
   (ii) brick on edge
   (iii) half brick thick
   (iv) one brick thick.

From the foregoing the following will apply.

Description 1 — 'foundations in trenches'
Description 2 — 'reduced to one brick thick'

Level 5 This level identifies the variables involved, such as size, colour and weight.

(a) For 'Concrete Work' such items as over 300 mm thick, not exceeding 150 mm thick.
(b) For 'Brickwork' such items as walls, skins of hollow walls, projections.

From the foregoing the following will apply.

Description 1 — 'not exceeding 150 mm thick'
Description 2 — 'projections'

The descriptions therefore are written as follows.

Description 1

CONCRETE WORK
PLAIN INSITU CONCRETE
NORMAL MIX 1:2:4/40 MM AGGREGATE
Foundations in trenches
   not exceeding 150 mm thick

Description 2

BRICKWORK AND BLOCKWORK
BRICKWORK
COMMON BRICKS IN CEMENT MORTAR (1:3)
Reduced to one brick thick
   projections

Very often certain portions of a description will repeat the contents of a preceding description. The traditional method of drafting such items is to use 'ditto' or 'do' to indicate the repetitive contents.

Example

Traditional Bill Format

		Qty	Rate	£
	PLAIN IN SITU CONCRETE			
	Concrete (1:2:4/40 mm aggregate)			
1	Foundations not exceeding 150 mm thick in trenches	m³		
2	Ditto exceeding 150 mm but not exceeding 300 mm thick do	m³		
3	Ditto exceeding 300 mm thick do	m³		
4	Beds 125 mm thick	m²		
5	Ditto 150 mm thick	m²		

Standard Description Bill Format

		Qty	Rate	£
	CONCRETE WORK			
	PLAIN IN SITU CONCRETE			
	NORMAL MIX 1:2:4/40 mm AGGREGATE			
	Foundations in trenches			
1	not exceeding 150 mm thick	m³		
2	exceeding 150 mm but not exceeding 300 mm thick	m³		
3	exceeding 300 mm thick	m³		
	Beds			
4	125 mm thick	m²		
5	150 mm thick	m²		

Standard descriptions may be considered to have the following advantages.

(a) The quality of written communications is improved by the consistency of the standard terminology.
(b) The estimator will gain certain advantages since the work is always described in the same terms.
(c) The processes involved in the drafting and interpretation of descriptions are simplified.
(d) Drafting the item descriptions can be left to junior technical staff.
(e) Editing procedures are eliminated and replaced by routine checks that can be applied by junior technical staff.
(f) Pricing data is presented in a simple form and not complicated by diverse styles in description composition.
(g) Tender examination procedures are improved by simplifying the ability to compare work/item prices and records.

It may be argued that the descriptions become stereotyped and lose identity in literary presentation. There is also a problem of producing a sufficient number of clauses to cover all eventualities. Certain non-standard items appear on occasions and difficulty may be experienced in dealing satisfactorily with these 'rogue' items.

Standard Phraseology

A major contribution to the standardisation of descriptions has been made by
Fletcher and Moore, *Standard Phraseology*.[4] This attempts to establish a
systematic approach to the composition of descriptions and serves as an
aide-mémoire to the complete requirements of the standard method of
measurement of building works, leading the user through a comprehensive check
list.

The development of computer techniques has provided an impetus towards
the use of standardised contract documents and many computer library
descriptions have adopted the 'standard phraseology' as a basis for computer
systems.

National Building Specification

Standard specifications are produced by using the National Building
Specification (N.B.S.).[5] The N.B.S. was first published in 1973 by National
Building Specification Ltd and is a library of standard specification clauses. A
subscription service was launched in 1975/6 to supersede the 1973 edition and
continuously update and expand technical content. The clauses are drafted in
such a way as to form optional statements that can be selected or rejected to
suit the needs of any particular project. Each clause represents one statement.
This means that for any one item to be specified a number of statements may
apply. The specifier merely selects the appropriate clauses that suit the item to
be specified. By selecting such clauses a comprehensive specification can be
produced with the minimum of writing and tailored to the needs of the project.
The N.B.S. is a library of standard specification clauses and not a standard
specification.

The library is divided into work sections that signify the kinds of finished
work such as: excavation and hardcore, brick/block walling, formwork for
concrete, structural timber. Each work section deals with one main type of
product in a defined range of applications. Each of the foregoing contains
clauses relating to (a) products and materials and (b) workmanship. The product
clauses are of two types.

(i) General clauses that specify the characteristics defining a range of
commodities and covering such requirements as physical properties,
storing, samples, approvals and testing. For example, the following may be
found under the heading of brick/block walling.
Samples: submit samples of each type of brick and block and obtain
approval before placing orders with suppliers.
Handling: unload and handle bricks and blocks without soiling, chipping or
causing other damage.
Storing: stack bricks and blocks on level hardstandings and protect from
inclement weather.

(ii) A specific item product clauses that (1) state the identity of the product, for example, calcium silicate bricks, clay common bricks, and (2) gives the qualitative statement about the product, for example, to BS 187 Part 2 Class 3A, to BS 3921 special quality.

Under the subheading of BRICKS/BLOCKS the following clauses will apply.

Calcium silicate bricks: to BS 187: Part 2, Class 3A
Clay common bricks: to BS 3921, special quality
Clay hollow blocks: to BS 3921 Section 2, keyed for plaster

The workmanship clauses cover instructions relating to the operations and contingencies that have an influence on the various parts of the work. For example

DRY WEATHER: wet clay bricks and blocks the minimum necessary to prevent mortar drying out prematurely.
COLD WEATHER: preheat water and sand for mortar as necessary to ensure a minimum temperature of 4 °C in the brickwork when laid.
UNIFORMITY: carry up work including both skins of cavity work with no portion more than 1.2 m above another at any time racking back between levels.
BRICKS: lay bricks on a full bed of mortar and fill all joints. Make bed and vertical joints of equal and consistent thickness.

Each of the clauses contained in the N.B.S. has a coded reference. The references are based on the CI/SfB classification, which is widely used for the classification of trade literature and product information. This means that any specification based on the N.B.S. establishes a more positive relationship with other documents such as drawings, quantities and trade literature. A specification based on the N.B.S. would therefore appear as follows.

F11   BRICK/BLOCK WALLING

SCHEDULE OF FINISHED WORK
F11:A   *BRICKWORK/BLOCKWORK TYPES*

	Brick/Block Type + Code	Mortar Group/Mix	Bond Type	Joints in Facework Type + Code
Type F11:A1	Calcium silicate Ff1.15	Group 3	Flemish	Weathered 4151
Type F11:A2	Clay common Fg2.10	Group 3	Flemish	Weathered 4151
Type F11:A3	Blocks Fg2.91	Group 5	Stretcher	Blockwork for plaster 4501

COMMODITIES

F11:A *GENERAL*
Aa0.01 The BSI documents referred to in this division are BS 187: Part 2: 1970 + amendment AMD 695 BS 3921: Part 2: 1969

F11:F *BRICKS/BLOCKS*
Fa0.01 Samples: submit samples of each type of brick and block and obtain approval before placing orders with suppliers.
Fa0.04 Handling: unload and handle bricks and blocks without soiling, chipping or other damage.
Fa0.06 Storing: stack bricks and blocks on level hardstanding and protect from inclement weather.

WORKMANSHIP

F11.1 *GENERAL*
1101 DRY WEATHER: wet clay bricks and blocks the minimum necessary to prevent mortar drying out prematurely.
1151 COLD WEATHER: preheat water and sand for mortar as necessary to ensure a minimum temperature of 4 °C in the brickwork when laid. Maintain the temperature of the work above freezing point until the mortar has fully hardened.

F11.2 *LAYING*
2251 UNIFORMITY: carry up work including both skins of cavity work with no portion more than 1.50 m above another at any time racking back between levels.
2351 BRICKS: lay bricks on a full bed of mortar and fill all joints. Make bed and vertical joints of equal and consistent thickness.

As a standard document the N.B.S. may be adopted for computer use. The standard specification clauses are stored in the computer with a code reference. This service is normally supplied through agencies who offer computer processing services and from whom a full text of the required clauses and a full job specification is obtained.

LIBRARIES OF DESCRIPTIONS

One of the main problems of data processing for quantity surveyors is description coverage. The basis of any system is a description library. In compiling such a library consideration must be given to the amount of description coverage. In so doing two conflicting requirements must be resolved. Firstly, the greatest possible coverage of likely bill descriptions must be examined. Secondly, an arrangement must be devised that is concise and compact and ensures a prompt location and identification of items. The description coverage should never be so limited as to create an uncontrollable number of items that are not covered by the library; nor should it be so full as to render the look-up time unacceptable.

One method used to give a desired result might be to write a program containing a library of all the possible items occurring in bills of quantities. Each item would have a unique coded reference for storage in memory. The problem in this instance is really a matter of the size of the library involved. The number of items needed to cover all the possible descriptions required by the various specifications is of such a magnitude as to preclude the use of this method. It has been estimated that well over 50 000 items would be needed to cover the basic requirements.

Libraries of descriptions may be considered under the two main headings

(a) standard library
(b) multiple library

Standard Library

Each item in the standard clause library represents a full description; an instance of this may be found in the N.B.S. This method facilitates the use of a numeric code that may be entered directly by the taker-off on his dimensions in the description column. More usually, this would be done by a coder after the taker-off has written his normal description.

Multiple Library

A bill description can be regarded as an amalgam of component phrases, as illustrated by standard descriptions. A greater description coverage is achieved, therefore, by writing a computer program containing a library of standard phrases that can be permutated to give the required description. A multiple library comprises skeleton descriptions or standard phrases and can thus be considered as (i) a library of skeleton descriptions, or (ii) an articulated phrase library.

A library of skeleton descriptions is composed of skeleton descriptions with a sublibrary of words and phrases that can be used to fill the gaps in the skeleton descriptions. By this method the skeleton framework of a description is used as a basis for the library with gaps for variables. Such variables as thickness of beds, depths of excavations, sizes of components and the like are picked out of a sublibrary and inserted in a relevant space in the skeleton description to complete the whole. In addition to the skelton phrases and words a library of skeleton headings and subheadings is provided for use in conjunction with the descriptions.

An articulated phrase library is composed of standard phrases that are so drafted as to be interchangeable and interconnected. The number of phrases that make up descriptions can vary considerably depending on the complexity of the work. In this way a greater amount of description coverage is achieved with the minimum amount of data. Each phrase is given a coded reference and a description is built up by combining the coded references in a predetermined order. One step towards rationalising the problem is the adoption of standard phraseology. Many systems in use have adopted the Fletcher/Moore phraseology and those derived by Monk and Dunstone, although the code structures for these libraries are very sophisticated.

CLASSIFICATION CATEGORIES

Another aspect of standardisation that affects the quantity surveyor's work is the development of classified categories relating to building elements and formats of bills of quantities.

Building Elements

A building element is a classification that may act as a basis for summarising costs and performance. It also provides a methodical approach to design, cost planning and comparison. Building elements are used more specifically as a basis for

(a) cost analysis
(b) approximate estimating
(c) procedure for taking-off
(d) bills of quantities format
(e) post-contract management purposes
(f) a design tool to provide alternative solutions to design problems.

Each element corresponds to a simple recognisable part of a building. It is possible to define an element as a section of work that corresponds to a design function — this means the part that forms the same function irrespective of its design and construction. The design functions such as walls, roofs and floors may be considered as building elements. External walls may be constructed of stone, brick or concrete; roofs may be constructed of concrete, timber or steel; each also has differing finishes.

The purpose of elemental classification is to provide a basis for the evaluation of factors affecting design in terms of cost and basic performance. A number of lists of elements has been prepared each of which has been developed to suit the needs of the user. For instance, a list prepared by the Technical Coordination Working Party at the Department of Education and Science[6] was influenced by the discipline imposed by an industrialised building system, mainly for schools.

There is no standardised list of elements recommended for general use. However, the following will indicate the typical elements used by the majority of quantity surveyors.

	List of Elements used by Architects Journal	List of Elements used by Building Cost Information Service
A	Preliminaries Contingencies	Preliminaries Contingencies
B	Work below lowest floor finish	Substructure

74

List of Elements used by Architects Journal	*List of Elements used by Building Cost Information Service*
C Frame	Frame
Upper floors	Upper floors
Roofs	Roofs
Roof lights	Stairs
Staircases	External walls
External walls	Windows and external doors
Windows	Internal walls and partitions
External doors	Internal doors
Internal structural walls	
Partitions	
Internal doors	
D Wall finishes	Wall finishes
Floor finishes	Floor finishes
Ceiling finishes	Ceiling finishes
Decorations	
E Fittings	Fittings and furnishings
F Sanitary fittings	Sanitary appliances
Waste soil and overflow pipes	Services equipment
Cold water services	Disposal installations
Hot water services	Water installations
Heating services	Heat source
Ventilation services	Space heating and air treatment
Gas services	Ventilating system
Electrical services	Gas installations
Special services	Lift and conveyor installations
	Protective installations
	Communication installations
	Special installations
	Builder's work in connection with services
	Builder's profit and attendance on services
G Drainage	Site works
External works	Drainage
	External services
	Minor building works

From the foregoing it will be seen that the elements may be grouped into primary divisions. Many elements are basically similar but some slight differences occur. For instance, 'finishes', 'windows' and 'external doors' may be included in the 'external walling' element, and 'finishes', 'internal doors' and 'partitions' may be included in the 'internal walling' element. Some clarification and definition is, therefore, necessary for each element used.

BILLS OF QUANTITIES FORMATS

It is usual to prepare a bill of quantities to serve the primary functions of obtaining competitive tenders on a uniform basis with the minimum of effort on the contractor's part. This document serves other functions as follows.

(a) It is usually incorporated as a contract document to form a schedule of rates. This creates a basis for the valuation of variations, which often occur during the progress of the works.
(b) It assists the quantity surveyor in making valuations of the progress of the works.
(c) It creates an itemised list of component parts and assists the contractor in assessing his requirements.
(d) It forms a source of cost data and affords a basis for cost analysis.
(e) It is a means of satisfying the client that he is paying a reasonable price.

The format or layout of the bills dictate their usefulness to the design team and contractor for pre- and post-contract needs. The traditional format for bills of quantities is based on a trade-by-trade presentation. This optimises the size of the bill and serves the primary function of obtaining tenders. Estimating is built up from the use of materials which shows a general uniformity to one trade. Trade groupings also help the contractor in obtaining subcontract quotations on a trade basis.

The traditional bill format does not fully utilise the information that has been prepared. Items in the bill are presented in a way that facilitates pricing for the purpose of tendering. Little consideration is given to the post-contract needs of the participants in the building process. For instance, the location of items is not apparent. The taker-off makes notes of the location of items in his headings or 'on waste'. This information is usually buried in the take-off and is not passed on to the succeeding stages of the design and construction processes. Measurement may also be carried out in sections that resemble some of the elements into which costs are later analysed. The order in the bill of quantities may not reflect this order of measurement. The contractor will need some or all of this information and will have to generate it afresh or call for a different bill of quantities arrangement. This service is only likely to be available if the bills have been produced by computer. In certain instances there is a complexity of measurement; the items as presented do not indicate the amount of labour, materials and plant involved. Units of measurement are consistently useful to the estimator or those responsible for organising resources. For instance, brickwork is measured as a superficial item and does not indicate the number of bricks, tonnes of sand or bags of cement. The estimator must evaluate the quantities of materials before he can arrive at a unit price. This should match the unit of measurement and he must then convert his figures. Later the contractor requires the number of bricks, tonnes of cement and sand for ordering purposes. This he does by reversing the calculations.

Certain attempts have been made to provide a format that is likely to give the fullest use of the project information prepared by the quantity surveyor. No single format provides information in a way that suits the needs of all concerned. The computer's ability to store, process and retrieve information provides an added facility for presenting data in a number of different formats from a single source, namely the taking-off. The arrangements of the constituent components of a bill of quantities into a specific sequence or format is known as 'sortation'.

The better-known bill formats developed in relation to the bill sortation are as follows.

(i) Elemental bills
(ii) Sectionalised trade bills
(iii) Activity bills
(iv) Operational bills

Elemental Bills

An elemental bill of quantities is one that is divided into suitable building elements instead of normal trade headings. The work within each elemental section, however, is billed in trade order. For example, the basic trades of 'excavator', 'concretor' and 'bricklayer' may give way to 'substructure', 'external walls' and 'internal partitions'. The latter would form the main sections of the bill with the basic trades as subsections. The object is to present the work in a manner that will enable more precise tendering and make the location of items more apparent and of use to all concerned.

While it may be easy to define a building element for the purpose of cost planning it is more difficult to divide a bill of quantities into elements complying with the strict definition. Certain elements may be indivisible and may combine several functions. For instance, the function of an external wall may be considered as (1) to keep out the weather, (2) a form of insulation both thermal and sound, (3) to support loads, dead, wind, floor and roofs, (4) to transmit light and ventilation (curtain walls). True comparisons depend on the functions and it may be necessary to refer to 'frame' and 'window' elements in order to arrive at a true indication of the wall's cost performance. An elemental bill involves a certain repetition of items. Where work is to be let to a subcontractor it is necessary to prepare an abstract of items before obtaining a quotation. The same problem applies when the estimator wishes to price a particular trade. He must look through the elements to gather all the items before he can judge the total quantities of materials. Repetition of items depends on the list of elements chosen.

Sectionalised Trade Bills

The sectionalised trade bill meets some of the contractor's objections to elemental bills and was developed by Hampshire County Council. The format is presented as a trade bill with elements as the main subdivisions. The whole is presented in a loose-leaf system with each element on a separate sheet. Unfortunately the estimator is still faced with a certain measure of repetition. Successful contractors are presented with several copies of the bill. This provides a choice of a trade order for buying materials and an elemental order for site use.

Activity Bills

An activity bill contains sections based on activity headings (operations) derived from a network analysis. This type of bill is measured in accordance with the standard method of measurement (S.M.M.) and prepared in the orthodox manner.

A network analysis is a prerequisite to determine the activities. Building

77

operations may be considered as being executed by a series of activities, each of which is interconnected. Some activities may be required to be completed before another begins. For instance, trenches need to be excavated before concrete foundations can be laid, and foundations need to be laid before brickwork can be built. A graphical representation consisting of a series of interconnected points is drawn. This establishes an interrelationship and sequence of activities and is known as a network analysis. While the network analysis for small jobs can be produced by most architects and quantity surveyors, for larger contracts it must be worked out in consultation with the contractor. This arrangement seems to rule out the activity bill for competitive tendering.

Operational Bills

An operational bill, like the activity bill, consists of sections with descriptions of operations which follow the building process. This type of bill was developed by the Building Research Establishment and concentrates on those building operations derived from a network analysis like the activity bill. The operations are more directly related to the manner in which the building costs are incurred.

The format of the bill departs from the accepted modes of measurement and provides information in a way that is of direct benefit to the contractor. Each operation is kept separate with materials shown separately from labour. The materials requirements are expressed in detail indicating the amounts of each commodity. The labour requirements are described in terms of the operations necessary for the construction of the building.

PROGRAMS FOR QUANTITY SURVEYORS

It is necessary to consider how electronic digital computers can best be programmed to perform the operation of supplying a service to quantity surveyors in a practical and economic manner. This will depend on the type of program written and the programmer's measure of the problem situation, which involves a knowledge and understanding of the requirements of quantity surveyors and a skill and experience in writing programs. It is important that the computer functions to serve the needs of the profession rather than vice versa. Certain constraints imposed by the idiosyncrasies of the computer might dictate a method of approach or presentation in format. However, with the right approach, realistic and useful results should be obtained. The right approach should involve the quantity surveyor, who, with a good general knowledge of computers, can assist the programmer in assessing the problem situation and get the requirements in perspective.

Programs may be considered under two main headings (a) single-purpose program and (b) general-purpose program.

A single-purpose program is one that is written for a singular purpose with a particular problem in mind. This type of program might be written by the computer user for such problems that relate to research and design and is often unsophisticated by professional standards.

General-purpose programs are written for particular types of situation and can be used repeatedly for solving problems of a similar nature. It is this type of program that is generally required by the quantity surveyor. These would be written by computer manufacturers and large computer owners.

The process of producing a bill of quantities involves a series of operations. For instance, the simple process of squaring dimensions involves the following procedures: (i) checking waste calculations, (ii) squaring dimensions, (iii) casting. Each operation is necessary to achieve the required results. In the same way a number of programs may be written, each performing an operation in a system of processing. A number of such related programs is known as a *suite* of programs. Programs written for use by quantity surveyors would in all probability comprise a suite. Many quantity surveyors make use of this type of program arrangement, known as *application packages*, which is discussed in more detail in chapter 11.

REFERENCES

1. ——*Standard Method of Measurement of Building Works* (Royal Institution of Chartered Surveyors and National Federation of Building Trades Employers, R.I.C.S., London, 1965).
2. —— *Code for the Measurement of Building Works in Small Dwellings* (Royal Institution of Chartered Surveyors and National Federation of Building Trades Employers, R.I.C.S., London, 1968).
3. ——*Standard Method of Measurement of Civil Engineering Quantities* (Institution of Civil Engineers, London, 1976).
4. L. Fletcher and T. Moore, *Standard Phraseology for Bills of Quantities* (George Godwin, London, 1965).
5. ——*National Building Specifications*, 4 vols (National Building Specifications, R.I.B.A., London).
6. ——*Building Industry Code* (Department of Education and Science, Technical Coordination Working Party, H.M.S.O., 1969).

8 BILLS OF QUANTITIES BY COMPUTER

At this juncture the effect of computerisation on the work of the quantity surveyor will be considered. A large proportion of the quantity surveyor's work is a tedious manual routine involving the speedy calculation and collation of items. Since the computer is very adept at performing these routines at a phenomenal speed the quantity surveyor should be attracted by this facility.

Since computers were first introduced there has been an absence of any standard approach, which has resulted in many and varied systems of processing. The quantity surveying profession in general remained uninterested in developing techniques. It has been left to several groups of surveyors with a pioneering spirit to continue with their isolated efforts. The reasons for this lack of enthusiasm were due to the uncertainty of the economics of a change to a system that was very expensive to develop, particularly with machines too small or too expensive for the work involved. At the same time pressure of work made it difficult for firms to release personnel to learn new techniques and to apply them. In consequence development has followed a number of paths and a confusing variety of systems has emerged. Although it is the absence of a standard approach that is the root of this divergence it is important to recognise that each problem is capable of diverse solution. A confusing variety of systems now confronts the quantity surveyor and it is only possible to consider the general characteristics of some of them in common use.

PROCEDURES

The systems developed for the preparation of bills of quantities by computer generally follow a similar pattern. Certain manual procedures have been supplanted by others that are necessary to execute the automated processes. The principal manual and computerised procedures for the preparation of bills of quantities are listed.

Procedures using Conventional Methods	Procedures using Cut and Shuffle	Procedures using Computers (figure 8.1)
1. Taking-off	1. Taking-off	1. Taking-off
2. Check waste calculations	2. Check waste calculations	2. Check waste calculations
3. Squaring and casting dimensions	3. Squaring and casting dimensions	3. Squaring and casting dimensions
4. Check squaring and casting	4. Check squaring and casting	4. Check squaring and casting
5. Abstracting	5. Slip sorting	5. Coding
6. Check abstracting	6. Collation of like items	6. Check coding

Procedures using Conventional Methods	Procedures using Cut and Shuffle	Procedures using Computers (figure 8.1)
7. Reduce totals	7. Total like items	7. Punching
8. Check reducing	8. Check totals	8. Check punching
9. Prepare draft bill	9. Reduce totals	9. Computer processing
10. Check draft bill	10. Check reducing	10. Check abstract/edit print-out
11. Edit draft bill	11. Edit	11. Prepare edit slips and input to machine
12. Type bill of quantities	12. Type bill of quantities	12. Produce bill of quantities

A comparison of both the manual and computer procedures shows that the processes of abstracting and draft billing have been supplanted with manual processes of coding and punching.

Taking-off

Taking-off follows the normal traditional procedures and changes very little, with the exception of the dimension sheets and possibly the order of take-off. One of the main features of computerised bills of quantities is the availability of different bill sortations. The first stage in the preparation of the bill is to decide the format required. The usual order of take-off for traditional bills using manual methods involves the measurement of sections of work or a trade-by-trade format using the Northern or London systems. This procedure has been affected when producing computerised bills since the take-off must be presented and referenced in sections commensurate with the bill sortations selected. This means that the take-off for an 'elemental' bill must contain items grouped into the elements chosen. Likewise, the location of items for a 'location' bill must be clearly indicated in the take-off. Where a number of bill sortations are required from one take-off, a suitable order and grouping of items must be adopted by the taker-off.

Each taker-off should be made familiar with the constraints imposed by the system he is using. A coding manual should be made available to eliminate any research into standard office bills for suitable descriptions and to act as a check-list for his work. He must also arrange his measuring to take advantage of any benefits the system may provide. This he is unable to do if he is insulated from the requirements of the code(s) used. However, the taker-off is still outside the automated process.

A typical dimension sheet is ruled to give a layout that is more acceptable for coding. The sheets are divided (a) vertically for dimensions, descriptions and coding, (b) horizontally to distinguish each item for coding purposes (see figure 8.2). Columns 1 to 4 inclusive are used for taking-off in the usual way. Column 5 is used for coding. The horizontal divisions, known as *lines*, create boxes into which item descriptions and coding are inserted. Taking-off follows the normal pattern with dimensions, timesing and squaring being inserted in the usual way. Neither dimensions nor descriptions are restricted to any one line and may be carried down to the next line as indicated. The formats of dimension sheets differ, depending on the system used.

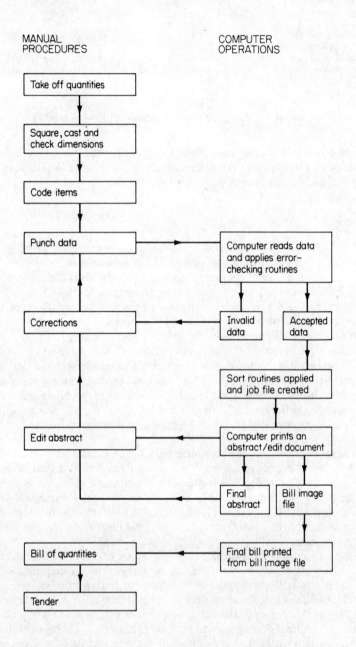

MANUAL
PROCEDURES

COMPUTER
OPERATIONS

Take off quantities

Square, cast and
check dimensions

Code items

Punch data

Computer reads data
and applies error–
checking routines

Corrections

Invalid
data

Accepted
data

Sort routines applied
and job file created

Edit abstract

Computer prints an
abstract/edit document

Final
abstract

Bill image
file

Bill of quantities

Final bill printed
from bill image file

Tender

Figure 8.1   Procedures for the preparation of bills of quantities

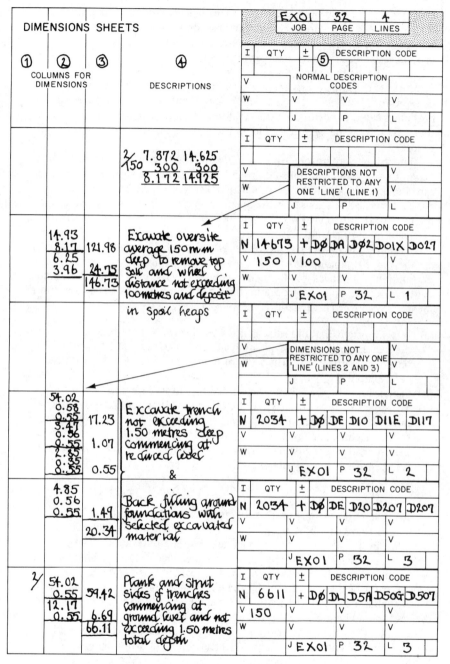

Figure 8.2   Example dimension sheet

83

Squaring and Casting

The computer should not be regarded as just a calculating machine since dimensions can be squared economically by comptometers and desk calculators. Its facility for squaring dimensions, therefore, is not always used and this procedure is considered by some quantity surveyors still to be a manual process. Each item in the taking-off may have a number of dimensions for calculating its quantity. If the dimensions are processed by computer each measurement must be reproduced in a punch coded form before input. This is time-consuming and nothing is saved in performing the operation. If dimensions are to be input in their raw state some benefit should accrue, like automatic coding[1] (see chapter 11).

Dimensions may now be squared more quickly than it takes to write them down, by using an electronic desk calculator. This device has the power and speed of a small or minicomputer and is designed to perform all the mathematical calculations that the quantity surveyor is likely to need. Some models can fit into a coat pocket and their cost should lie within the financial means of most organisations.

Computer programs may also be simplified if the dimensions are squared and cast before they are input to the computer. This means that there will be only one amount for a measured item. Many quantity surveyors prefer to maintain control over the squaring procedure for a number of reasons apart from those mentioned above. These are listed as follows.

(a) It is important that bill editing at the final stage be kept to a minimum and the manual procedures of squaring and checking assist in a technical edit before data processing. The work of juniors can be properly scrutinised before input.
(b) Dimensions may be suitably annotated for future reference.
(c) The quantity surveyor may require certain information from the original take-off for post-contract use. Dimensions squared by the computer are lost to the machine and are difficult to recover for such uses.
(d) Less punching is required, thereby reducing the error zone. Correcting errors is expensive, being a waste of computer power.

Coding

Coding is an extra manual process that is necessary to prepare data when using computers to prepare bills of quantities. It is concerned with the translation of the items in the take-off into terms suitable for computer input. The process does not necessarily require any technical knowledge of the machine but does require a certain skill in transposing the take-off data into coded terms. This process must be distinguished from the coding necessary for data coordination.

To enable the descriptive matter to be collated and sorted by the computer the data should be presented to the machine in a standardised form. This is achieved by the use of a description code that provides a structure for bill descriptions and enables the computer to carry out the working-up stage in the production of bills of quantities. The description code comprises a standard library of coded descriptions. In some systems a comprehensive standard library is designed to give a full coverage to enable work to be measured in accordance with the Standard Method of Measurement (S.M.M.). At the other end of the scale a *library-less* system is used where descriptions are derived from an *ad hoc* library compiled for each job. When compiling a standard library of descriptions for computer use the following should be considered: (a) the composition of the descriptions and (b) the structure of the associated code. It should also possess the following characteristics.[2]

(i) The format of the standard library should be so constructed as to reach the optimum in size and reference time. The library should be as comprehensive as is practicable and yet be sufficiently flexible to allow for changes in building materials and techniques. It is important that the look-up time be reduced to a minimum.

(ii) The library should be based on a close analysis of the methods of measurement used. In the case of building contracts this is the S.M.M. Separate libraries may be necessary for each code of measurement.

(iii) Descriptions should be restricted to meaningful entities, and if the library is of an 'articulated' nature each code facet should represent a phrase or articulation of the description.

(iv) The phraseology must be grouped according to pricing principles. The basic structure of the library should be amenable to standard pricing and should have an ability to link up with an automated system of pricing and should provide information in respect of labour, materials and plant as well as statistical information.

(v) A certain rationalisation should be maintained to conform with any requirements of coordinated information systems. Standardised descriptions should be used wherever possible.

(vi) The code structure should be designed to reduce the look-up time to a minimum and should contain an easily identifiable classification pattern. There should be very little searching for code elements and all facets should be clearly shown. Simple identifiable references are easier to handle.

An important factor to be considered when compiling a standard library is storage space, not only in the code manual but also in the computer. There is a greater power with an articulated phrase usage. The number of different descriptions will be increased if phrase storage is provided for different levels of a description. Assume that there are eight, twelve and fourteen phrases available in three levels. The number of descriptions available, therefore, will be $8 \times 12 \times 14 = 1344$; the number of lines required for storage will only be $8 + 12 + 14 = 34$. In this manner the scope and coverage of the library is increased by permutation.

The coding process is usually carried out by the quantity surveyor who may employ takers-off or other assistants with a knowledge of the code disciplines of the library and the system in use. A number of experienced coders are employed by companies providing computer services who may code their customers' work. However, they normally supplement the efforts of their customers when they (the customers) are busy. Nevertheless their policy is usually to encourage users to carry out their own coding. If the quantity surveyor is to remain responsible for the finished document he must maintain control over the coding procedure.

It is incumbent upon the quantity surveyor to check the coding if it is carried out by an external agency. He must ensure the accuracy of the finished product or he may be in breach of his duty to his client. It is at the coding stage that bill descriptions are determined and responsibility for these should remain in the hands of the quantity surveyor.

The production of bills of quantities by computer should save time in the working-up process but it has been found that it can be a time-consuming and tedious task. Very little time may be saved, especially if large numbers of items appear that are not in the standard library. The coding procedure varies from office to office but the best results are usually obtained where the takers-off code their own dimensions. Some offices consider that clerical assistants can code efficiently under supervision after two or three months practice.

The code fields set out in column 5 of figure 8.2 contain all the coded information necessary for the computer to process an item. This involves the following information (see figure 8.2).

(1) The type of input, which may be normal dimensions, short code dimensions or unit quantities (I).
(2) A quantity field containing the quantity of the item. This is normally expressed to two places of decimals (QTY).
(3) An indication of the arithmetic value of the item indicating whether it is an addition (+) or subtraction (−).
(4) A code field for the description code. This may be divided into a number of fixed fields for the coding of various levels of an articulated phrase description (DESCRIPTION CODE).
(5) Code fields, which provide for variable inserts. These are usually found in level 5 of a description and may consist of dimensions or alphanumeric information (V).
(6) A reference for a description that is to be written short to ensure that it appears in the correct position in the bill of quantities. Variable inserts may also be required for these items (W) and (V).
(7) A reference indicating an 'in number' quantity of certain items that is required to be specified as part of the description.
(8) The job reference, which must be included as part of the code of each item (J).
(9) The page and line reference indicating the sequence of the item in the take-off. This also acts as a check on the number of items. For example, page 32 of the dimension sheets has four lines (4 items): the numbering of the items would be 32 4 1 to 32 4 4. In this way the computer can check that the correct number of items have been inserted as input (P) and (L).

To help eliminate errors in interpreting the coding certain handwriting conventions must be observed. These may vary slightly with different offices. The following, however, are some typical examples.

Letter O — written $\emptyset$ to distinguish it from the number 0 (zero)
Letter I — written I to distinguish it from the number 1

Letter Z – written Ƶ to distinguish it from the number 2
Letter S – written $ to distinguish it from the number 5

In addition to the above conventions care must be taken over the following pairs of letters and figures.

G and 9
7 and Y
Y and Z

Great care must be exercise to ensure clear handwriting, otherwise errors will be produced by punch operators who have misread the handwriting.

The procedure consists of translating the descriptions into code by reference to a standard coded library of descriptions. The codes are recorded in the appropriate boxes on the dimension sheets. Each coding box is usually divided into code fields, which may be subdivided into columns where the relevant code references are placed (see figure 8.2, column 5). Coding every item can be a long and tiresome operation. However, certain items at particular levels of a description may have a common code that simplifies the task. Certain systems[3] incorporate a routine providing a *carry-down* facility; this eliminates the need to insert code references where there is a level of description immediately preceding the item with a common code reference at that level. This means that the common references may be left blank where there are successive common code references (figure 8.2, line 3). The computer will pick up the codes used on the items immediately preceding it and carry down the references where blanks occur. The facility only applies to those items on a single dimension sheet; it will not work from page to page and only applies when items follow each other on any one page; the first item on every page must therefore possess the code reference in full.

Checking Processes

It is essential that the various procedures maintain routine checks. Although error-checking routines are incorporated in programs this does not preclude the need for manual checks on such processes as taking-off, coding and punching. The accuracy of the finished product depends on the accuracy of the input data. This involves checking information before it is input to the computer. The computer will check to ensure that inserts of dimensions and cash amounts of prime cost and provisional sums are correctly received; checks will also be made to ensure that codes are complete in all respects. It will not pick up incorrectly coded items but only errors of syntax. A manual check of coded items is therefore necessary. For instance, the input may be 112.56 in cubic metres, square metres or linear metres. The computer will read a measurement in a code field and assume the appropriate unit according to the description associated with the measurement. The computer will read 112.56 m^3 if the description is concrete foundations, 112.56 m^2 if brickwork, and 112.56 m if timber joists. Not only must a check be made on the correct unit of measurement but also that the numeric characters are in the correct column. The decimal point is inferred and not recorded as input.

Punching

Data input are usually punched paper tape or punched card. The procedure involves the transfer of coded data from the dimension sheets on to the input media. Punching is left in the hands of trained and experienced operators and if the quantity surveyor wishes to punch his own data he will have to employ specially trained staff who may be required to work solely on this operation. In this way he will have the advantage of keeping the dimensions in his office. There are certain disadvantages that may preclude the quantity surveyor performing this operation.

(a) The capital equipment needed must be purchased or hired and a number of machines may be necessary to avoid problems when breakdowns occur.
(b) Maintenance of the machines has to be arranged and its cost covered.
(c) A large punch-operating section may be needed if serious bottlenecks are to be avoided at peak periods. Slack periods in this case will be expensive. Punching may, however, be contracted out.

Verification of Punched Paper Tape

The accuracy of punching depends on the operator and a second operation must be introduced to avoid errors. This is known as 'verification'. The verification of punching may be carried out in one of the following ways.[4]

(a) *The Call-over Method* This method involves the use of a tape punch combined with a printer. A printed copy is made at the same time as the data are being punched, and this may then be checked with the original as a verification of the accuracy of the punched tape. A disadvantage of this method is that when checking by sight it is easy to miss an error.

(b) *Two-tape Method* This method, which is probably the most widely used form of verification, is carried out by a tape verifier consisting of a tape reader, tape punch and keyboard. A tape is punched by an operator using an ordinary punch machine. A second tape is prepared by feeding the tape initially prepared into a tape verifier. A second operator produces a second tape after a key has been depressed and the character compared with that on the initial tape. If the two characters agree the tape is punched and moves on to the next character. If the characters do not agree the keyboard will lock, thus preventing the character being punched. The error must then be checked with the source document and the correct character punched.

(c) *Three-tape Method* This method initially involves the preparation of two tapes by different operators referring to the same source material. Verification is then carried out by feeding the two tapes into a verifier, which reads them simultaneously and compares them character by character. If they agree the character is automatically punched on to a third tape. If they do not agree the machine will stop and allow the error to be corrected.

88

Verification of Punched Cards

The verification of punched cards is performed by a second operator who punches the cards a second time from the same source document. The machine used for this operation checks that the holes have been punched in the correct positions, by means of an electrical sensing device. If holes already exist in the correct positions, the card is released. Where no hole exists the operator's attention is drawn to the difference. A check must then be made for errors and the card repunched if necessary.

An alternative method involves the use of a machine that alters the shape of the hole. A round hole originally punched is converted into an oval one for verification. The test is for the presence of round holes which indicate punching errors.

Any cards that need to be repunched must be verified in their turn to ensure the accuracy of the correction. The accuracy of punching depends on the training and experience of the operator and the legibility of the source information. A good operator can produce something in the region of 300 punched cards per hour with errors of less than 1 per cent.

COMPUTER PROCESSING

The main problem when producing bills of quantities lies with the method of processing the data. It is in this process that the computer can be of greatest help.

The particular system used for processing will depend on the hardware available. This usually comprises a general purpose digital computer with a configuration consisting of a core storage with a capacity that is sufficient to hold (a) the executive or compiler, (b) any program(s) necessary for that particular stage in the processing, (c) the raw data requiring to be processed and (d) space for processing the data. This is supplemented by auxiliary storage in the form of peripheral file-holding devices, which may be used for recording the results of processing.

Although the core storage may be adequate for the purposes of retaining the programs and performing the processing operations the system may be seriously impeded if the auxiliary storage is inadequate. The configuration must therefore be equal to the needs and dictates of the program(s) written for the routines to be performed.

Programs are written with the object of performing specific routines in the processing of data for the production of bills of quantities. Each routine forms a definite step in the system of processing. Routines are written to record, sort, merge, collate or print data, and the application of programs, each of which performs a specific task in a set sequence, forms a computerised system of processing. A suite of programs forming a complete system of processing is known as a *package*. These may be purchased from certain organisations who have developed the systems for use on a commercial basis.

The general characteristics of a system[5,6] for producing bills of quantities may be described as follows (see figure 8.3).

(i) The punched paper tape or punched cards containing the coded data from the taking-off are fed into a tape reader or card reader on line to the

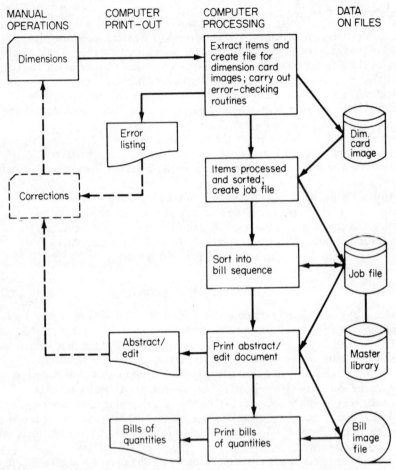

MANUAL OPERATIONS	COMPUTER PRINT-OUT	COMPUTER PROCESSING	DATA ON FILES

Dimensions

Extract items and create file for dimension card images; carry out error-checking routines

Dim. card image

Error listing

Corrections

Items processed and sorted; create job file

Sort into bill sequence

Job file

Abstract/ edit

Print abstract/ edit document

Master library

Bills of quantities

Print bills of quantities

Bill image file

Figure 8.3   System flowchart for the preparation of bills of quantities by computer

computer. The data are read and transferred for processing, which is carried out automatically. If dimensions have not been squared the computer will perform this operation and produce the results as output.

(ii) The computer reads the input data on to a file-holding device, thus creating a dimension image file on magnetic tape or disc. The coded information is transferred in the order recorded in the take-off. Each item in the take-off forms a record in the computer's memory. The file thus contains individual records of items of taking-off, each consisting of a measurement and description.

(iii) Checking routines are written into programs to reduce computer time that would be wasted in handling invalid data. The computer will pick out those errors that can be detected by a logical examination of the take-off at its face value. Errors detected at this stage may comprise invalid library codes or missing variables or records and the like. For instance, the check would reveal such errors as a missing sign indicating addition or subtraction, an entry in the quantity field that is not numeric, or any invalid entries in various other fields. Scrutiny is also made of every record to ensure that every line has been coded. The computer also checks that no pages are missing or page numbers duplicated. It performs a recheck on all records to make sure that each one has all the entries, such as variables, it will require for the bill. All errors so detected by this check are output for correction. The output from the computer is a list of such errors indicating the page and line number of each. Errors may then easily be traced back to the original take-off. Corrections are made by repunching. Error messages are produced on the error listing, which may be translated by reference to an error message book. This contains a list of each type of error with an appropriate error message and the action to be taken to correct it. The computer in some instances will make assumptions when certain errors appear. For instance, the required arithmetic sign may be missing. In this case the quantity will be assumed to be always addition and the sign to be plus (+); unless the sign is required to be minus (−), indicating a subtraction, no correction need be made.

(iv) The programs are rerun until the error listings are clear and all corrections are found to have the desired result. This means that the next stage of the processing can proceed.

(v) The processing is continued by applying programs with sort routines, which perform the abstracting process and produce an *abstract/edit* print-out. This provides the quantity surveyor with an opportunity of checking the full text of each item in bill order. It also provides the essential link between the take-off and the bill of quantities. The format of this document varies with each system used and very often a knowledge of the discipline of the standard library and of the system itself is essential to read the document. However, once these have been mastered the information can be gleaned very easily.

(vi) The texts of the bill items are built up by the computer by reference of the codes to the standard library of descriptions or master library file; the code is examined and the appropriate library item selected. The master library file is stored on magnetic disc for ease of reference. The routines provided by the programs will merge like items, insert non-standard items, summarise and resort as necessary. Notwithstanding the error-checking routines in the earlier stages of processing, it is necessary to maintain a check throughout. The abstract/edit print-out may also show errors that will be highlighted. The error messages may indicate such discrepancies as total quantities indicated as nil or minus, codes not found on the library, or a missing item for a written short code. These types of error are more of a technical nature, which go a little deeper than the syntax errors check in earlier routines.

(vii) A sequential *bill image* file containing the results of processing as shown on the abstract/edit print-out is created on magnetic tape or disc. From this file the final bill can be produced.

Editing

The abstract/edit print-out is checked by the quantity surveyor. This is a technical edit and any necessary changes can still be made. Many alterations may be made on large contracts that are not necessarily mistakes. Certain items may have their codes modified or rogue descriptions rephrased because sufficient information is not given or the description does not convey the full extent and character of the work. The best practice is to read through the whole print-out first and merely insert crosses in coloured pencil beside those items that need correcting. This produces an edit that does not disturb the surveyor's train of thought. Details of the corrections may be inserted on a second reading of the document. Great care is necessary when editing and correcting and it is considered good practice to alter the coding on the take-off as corrections are made.

It is at this stage that the refinement is applied, and a check is necessary on the effect of any alterations in producing a satisfactory bill. A rerun therefore is made and a second print-out obtained. A few errors are inevitably found on this document, which must be corrected. Once these have been rectified the final stage can be performed and a print-out of the final bill obtained.

Editing corrections should be marked on the draft abstract in a distinctive coloured ink so that they can be picked out easily. Corrections are made by preparing amendment cards or tape verified in the usual manner. These are then input to the computer and merged with the other data on the bill image file.

Records may be corrected in a number of ways, as follows.

(1) If punched cards are used a correction may be made by repunching a card and replacing the erroneous card with the correct one. This method is used only when a large number of punching or coding errors occur. The job file comprises the tray of cards that is fed into the computer.

(2) A record may be corrected by inserting a deduct item with an identical code and quantity to the incorrect item. The total quantity of that item will then be made zero and the item ignored by the computer on the final document. A record is created by inserting the correct version of the item, which will supersede the erroneous record. Care must be taken to ensure that the new item inserted to cancel the error is coded exactly as the original with the exception of page and line numbers.

(3) An existing record may be amended or deleted by the use of a program written to perform this routine. The method is quick and simple and only requires the insertion of coded corrections. The computer will pick up the corrections and make the necessary adjustments on the job file.

One of the most important considerations when updating the job file is the result of the carry-down facility. If certain items have been incorrectly coded it may have an effect on the items on the dimension sheets immediately following it. All the items that follow on the same dimension sheet must be carefully checked.

Production of Bills of Quantities

The final stage of processing is the print-out of the bill of quantities. The computer prints the bill from the bill image file. A program is used that applies

			£
DO		EXCAVATION AND EARTHWORK		
DA		SITE PREPARATION		
D02		PRESERVING VEGETABLE SOIL		
D01X		EXCAVATING		
D027	1	AVERAGE 150 MM DEEP ; DEPOSITING IN PERMANENT SPOIL HEAPS AVERAGE 100 M FROM EXCAVATION ; ON SITE	201	M2
D027	2	AVERAGE 200 MM DEEP ; DEPOSITING IN PERMANENT SPOIL HEAPS AVERAGE 100 M FROM EXCAVATION ; ON SITE	90	M2
DE		EXCAVATION		
D10		EXCAVATING		
D10G		SURFACES TO REDUCE LEVELS		
D10P	3	AVERAGE 150 MM DEEP	50	M2
D11E		TRENCHES ; STARTING FROM REDUCED LEVEL		
D117	4	NOT EXCEEDING 1.500 M DEEP	29	M3
D20		DISPOSAL		
D23X		SURPLUS EXCAVATED MATERIAL		
D207	5	BACKFILLING	20	M3
D24G	6	SPREADING ON SITE AVERAGE 100 M FROM EXCAVATION	17	M3
D30		SURFACE TREATMENTS		
D321		SURFACES OF GROUND		
D303	7	LEVELLING ; COMPACTING	55	M2

PAGE NO   9   TO COLLECTION   £

Figure 8.4   Example print-out of bills of quantities

93

a routine to sort automatically the bill image file into the required bill format with bill references, headings, page numbering and quantities and descriptions in their respective columns (see figure 8.4). It is difficult to determine in advance the number of lines that a description will occupy or to allocate the appropriate item reference. These decisions are left to the computer as dictated by the program. An image of the bill may be stored indefinitely on magnetic tape.

The bill sortation is selected by inserting a 'lead card', which contains all the information relating to the format required. This supplies the vital references necessary to select the appropriate programs. Copies of the bill may be produced by the computer but some surveyors prefer to reproduce the bills by obtaining photostat copies of the print-out. Others prefer to reproduce typewritten copies from the print-out.

ROGUE ITEMS

The problem of compiling a library of standard descriptions is to arrive at the number of standard items necessary to produce a bill of quantities. The simpler the system of coding the more items are likely to be required. It is therefore necessary to provide a routine in the program to cater for the non-standard items that are required. These non-standard items are known as *rogue items*. A rogue item may then be defined as any phrase or description that does not appear in the standard library and yet requires a code. Spaces are usually left between codes at all levels to allow for the future additions to the library filled by rogue items. Rogue descriptions and standard descriptions can be combined when coding any one description.

It is necessary to prepare a rogue description library to act as an extension of the standard library of descriptions. This will contain the non-standard or rogue items with code references that closely relate to the codes in the standard library. The two libraries then are read together when the computer compiles the job file. As an example assume that the following item is required in a bill of quantities.

> M2   Red rustic facing brick from Supabrick Co.; in cement mortar (1:3); one brick thick; weather struck pointing both sides; walls; Flemish bond

The rogue is a level 3 articulated phrase, which is

> Red rustic facing brick from Supabrick Co.; in cement mortar (1:3)

A rogue library is compiled by preparing rogue description sheets containing the following information.[3,7]

(1) A reference to the type of input; assume this to be 'R' indicating a rogue input.
(2) A reference to the type of insertion; this may be an addition or a deletion from the library. Assume this to be 'A' for addition and 'D' for deletion.
(3) The level of the rogue in the description; this may be at level 1, 2, 3, 4 or 5. This is '3' for the example set.
(4) The code of the rogue description; when choosing an appropriate code for the rogue description some consideration must be given to the position of

the rogue in the bill. A code should be chosen that will bring it into the desired position in relation to the other items. The code references in the standard library are usually sequential with spaces to allow the selection of codes for rogue items to maintain a sequential order. The rogue description created for any job may be assumed to be an extension of the standard library. Assume the code to be 'G51'.

(5) The unit of measurement; this is 'M2' for the example given.

(6) The rogue description; this is written exactly as it is required to be printed in the bill.

(7) If the description is so long that it needs more than one punched card a reference is usually placed in the last column to keep the series of cards in the right sequence. This is usually a numeric or alphabetic sequential reference.

The coding of the item quoted as an example would be coded for input as follows.

1	2	3	4	5	6	7
R	A	3	G51	M2	SAND FACED FACING BRICK FROM SUPABRICK CO IN CEM	A
R	A	3	G51	M2	ENT MORTAR (1:3)	B

Note   The above example indicates how a description would be entered on to two cards for input; the code field usually extends to eighty columns on card input and two cards only are required if the rogue description and its associated code contains more than seventy-nine characters.

The unique code reference 'G51' (column 4 above) would also appear in the normal dimensions code for the item of brickwork used as the example. This would be a level 3 code for the description.

The description code therefore would appear as follows.

GØ G6 G51 G46A G600

Level	Code	Description
1	GØ	Brickwork and Blockwork
2	G6	Brickwork entirely of facings
(*)3	G51	Sand faced facing brick from Supabrick Co; in cement mortar (1:3)
4	G46A	One brick thick; weather struck pointing both sides
5	G600	Walls; FLEMISH bond

(*) indicates the position of the rogue in the description code

The descriptions at levels 1, 2, 4 and 5 would be obtained by reference to the standard library of descriptions; the rogue description at level 3 would be obtained by reference to the rogue description library.

Rogue headings and subheadings are treated in the same manner as rogue descriptions; these are suitably coded and entered on the rogue library.

95

SHORT CODING

The most frequently occurring problems in connection with coding are the lengths of certain codes and the time taken to perform the coding process. A balance must be maintained between the long codes necessary to provide an adequate description coverage and the short codes desirable for practical use. A technique known as *short coding* has been developed, which reduces the lengths of codes and economically curtails the coding process.[3,8] The technique makes use of the *job files* prepared when processing bills of quantities.

A short code library is required, which is created by a computer program using an existing job file that has been updated with bill references during the production of a bill of quantities. The library may be set up and updated as shown in figure 8.5. The contents of the library consist of the library codes, and variables for those items as set up for previous bills processed by computer. Each item is allocated a short code reference, which is translated by the computer into a full description code via the short code library. A full description is generated from a shorter form of coding. For instance, the short code reference for

'102 MM x 150 MM Precast reinforced concrete lintel with one 12 MM bar' (figure 8.4)

may be '0709' (item number), assuming the contents of the short code library are based on the job file for a bill of quantities. Short code references are created by the numerical sequence (item numbers, see figure 8.4) established in the short code library (see figure 8.5).

Example   Assume the item required is

8M   102 MM x 150 MM Precast reinforced concrete lintel
.... one 12 MM bar

The code reference will be

(1) The nature of input (S)
(2) The quantity (8.00 M)
(3) The arithmetic value (+)
(4) The short code (0709)

The description code therefore will read as follows

'S 800 + 0709'

A bill that contains as many frequently recurring items as possible should be selected in order to deploy the short code facility effectively. In order to achieve a reasonably comprehensive coverage it may be necessary to update the library from time to time and incorporate new items. Programs have been written for this purpose. This means that two or three job files can be merged to produce a master short code library.

No matter how careful a choice is made of bills of quantities there will be several items not in the library. It is usually found that after three revisions of the library the number of new items are so few that it is uneconomical to incorporate them.

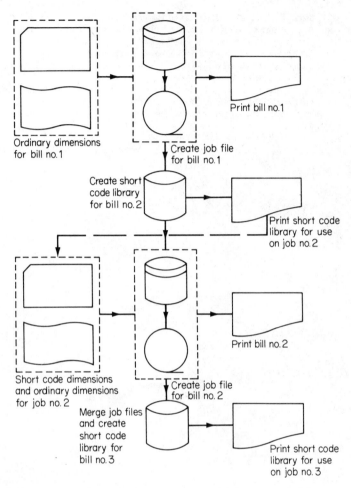

Ordinary dimensions
for bill no.1

Create job file
for bill no.1

Create short
code library
for bill no.2

Print bill no.1

Print short code
library for use
on job no.2

Short code dimensions
and ordinary dimensions
for job no.2

Create job file
for bill no.2

Merge job files
and create
short code
library for
bill no.3

Print bill no.2

Print short code
library for use
on job no.3

Figure 8.5   Flowchart for the creation of a short code library

The preparation of bills of quantities appears to be one of the principal uses the quantity surveyor makes of the computer. Although it has been in use for a number of years the computer has still to be proved in the eyes of many quantity surveyors. However, its use in a number of offices has provided a more accurate and comprehensive service. The more a computer is used the more its usefulness becomes apparent. The successive use of short coding is an effective way of achieving this objective.

REFERENCES

1. —— *The DG System: Computer Techniques Billing Systems* (Development Group of Chartered Quantity Surveyors, R.I.C.S., London, 1973).

2. —— *Computer Techniques* (Research and Information Group of the Quantity Surveyors' Committee of the Royal Institution of Chartered Surveyors, R.I.C.S., London).
3. T. F. Fry, *Computer Appreciation* (Butterworth, London, 1970).
4. —— *Computer System Handbook*, Part 1M, Consortium of Local Authorities Special Programme: Central Development Group.
5. —— *Computer System Handbook*, Part 2M, Consortium of Local Authorities Special Programme: Central Development Group.
6. —— *Computer System General Introduction*, Local Authorities Management Services and Computer Committee: Steering and Development Group.
7. R. Swanston, 'CLASP Computer System', *Chart. Surv.*, 102 (1970) pp. 411–16.
8. K. W. Monk and T. D. Dunstone, 'Short Coding in Practical Surveying by Computer', *Chart. Surv.*, 99 (1967) pp. 420–2.

9 UNIT QUANTITIES

The use of unit quantities is a facility that has been devised with the aid of the computer to simplify taking-off and to save time in coding.[1,2,3,4] Quantities are often required from standard components and the quantity surveyor is faced with the tiresome task of repeatedly having to measure the same items with the same dimensions.

One of the main benefits that can be derived from the use of a computer is its ability to store large quantities of information with a rapid retrieval facility. This facility has been utilised to provide benefit to the quantity surveyor in the preparation of bills of quantities. The basic principle is to measure the items of standard components of a building and to store them in the computer's memory bank. When the quantity surveyor requires this component or a number of components on a subsequent job, all that is required is a code for the unit quantities. By reference to this code the computer will generate the quantities relative to the number of components measured and will incorporate them in a bill of quantities.

A unit quantity library is required containing details of the components. Each section of the library should cover one standard component, which means that the library can be produced in sections as the details are made available. Although the components are considered to be standard there may be certain variables such as location of the component, differences in the specification of the parts and finishes or sizes of the items. This means that a section of the unit quantity library giving details of a standard component may contain a number of variable details.

The quantity surveyor, when preparing unit quantities, must first familiarise himself with every item involved and with their relationship with each other. Every possible situation and the ways that the different situations may affect the component as a whole must be examined. A set of quantities is required for each separate part of the component in every possible situation in which the component may be used. This is achieved by examining the variables, known as *parameters* and measuring the component as it is affected by that situation. The task of preparing unit quantities for all the possible variations would become uneconomic if the whole component were taken off for each possible situation.

PROCEDURE

The procedure to be adopted for the preparation of a unit quantity library is as follows.

(a) Choose the Component to be Incorporated into the Library

This is usually a standard unit or detail in common use. For the purpose of illustration the store unit shown on figure 9.1 is used as an example.

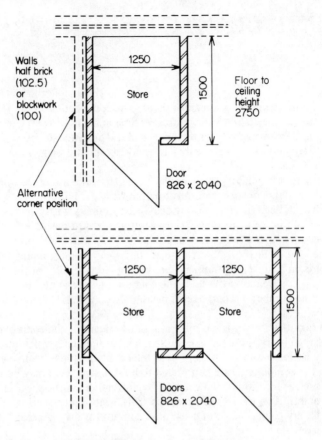

Figure 9.1   Unit quantities — single and double store units

(b) Study the Drawings

This is a prerequisite to taking-off. Having decided on a component the problem
is to break it down into items that will suit the varying situations that may
affect the component as a whole. An individual item is required for each
separate part of the component for every possible situation. The parameters and
range of items must be determined with the study of the drawings.

Specification Notes
    Walls
    Brickwork: to be common bricks in cement mortar (1:3)
    Blockwork: to be solid concrete blocks to BS 2028 in cement mortar (1:3)
    Finishing
    External face: render and set in gypsum plaster; apply three coats
        emulsion paint to plastered surfaces
    Internal face: form fair face and flush pointing to brick walls and

partitions; render in cement and sand (1:3) to blockwork walls and
partitions

Doors

Fit 44 mm thick flush doors size 826 x 2040 mm with skeleton core and
covered both sides with plywood including hardwood lipping to
stiles

Fit 100 x 75 mm softwood rebated and moulded door frame

Fit rim lock (reference RL 123)

Paint woodwork with two undercoats and one finishing coat full gloss
finish

(c) Determine the Parameters based on the Variables

The example is based on a store with the following parameters.

(i) The store may comprise a single or double unit.
(ii) The store may be in a corner position or on an open wall.
(iii) The store may be constructed of common bricks or blockwork.

If the following notations are assumed the range of situations can be assessed.

Parameter 1: S = Single unit
D = Double unit

Parameter 2: C = Corner situation
O = Open wall situation ($\emptyset$ in figures)

Parameter 3: CB = Built in common bricks
BL = Built in blockwork

With three parameters each of which has two possibilities the number of
combinations will be 2 x 2 x 2 = 8 as follows.

Possibilities	Parameter		
	1	2	3
(i)	S	O	CB
(ii)	S	C	CB
(iii)	S	O	BL
(iv)	S	C	BL
(v)	D	O	CB
(vi)	D	C	CB
(vii)	D	O	BL
(viii)	D	C	BL

(d) Take Off 'Preliminary' Unit Quantities

These are based on a particular combination of parameters which gives as wide a
range of items as possible. The example taken assumes the store is a single unit
on an open wall and constructed of common bricks (S O CB) (see figure 9.2).
This establishes the items to be incorporated into the unit quantity library and
acts as a guide when taking off the skeleton unit quantities. All descriptions
should be based on the Library of Standard Descriptions and rogue items should
be reduced to a minimum. Difficulties arise every time a rogue item is used since

			I	QTY	±	DESCRIPTION CODE	
		Preliminary unit quantities (1) Single unit in an open position with Common bricks (ref.S∅CB)					
			V		V	V	V
			W		V	V	V
			E	F	J	P	L
4.25 2.75		2/1500 3.000 1.250 4.250 Half Brick wall in common bricks in cement mortar (1:3) with four face one side	I	QTY	±	DESCRIPTION CODE	
			V		V	V	V
			W		V	V	V
			E	F	J	P	L
0.96 2.18		826 Fr 75 2.040 2/65 130 Reb 10 65 956 65 2.105 Lint 75 (door) 2.180 Ddt do.	I	QTY	±	DESCRIPTION CODE	
			V		V	V	V
			W		V	V	V
			E	F	J	P	L
1.16		956 2/103 206 1.162 102x75 precast concrete lintel reinforced with one 10mm bar	I	QTY	±	DESCRIPTION CODE	
			V		V	V	V
			W		V	V	V
			E	F	J	P	L
4.46 2.75		1 250 2/103 206 1.456 2/500 3.000 R&S walls in 4.456 gypsum plaster	I	QTY	±	DESCRIPTION CODE	
			V		V	V	V
			W		V	V	V
			E	F	J	P	L
		& Three coats emulsion to plastered walls (intly)	I	QTY	±	DESCRIPTION CODE	
			V		V	V	V
			W		V	V	V
			E	F	J	P	L
0.96 2.11		Ddt fust two items	I	QTY	±	DESCRIPTION CODE	
			V		V	V	V
			W		V	V	V
			E	F	J	P	L

Figure 9.2.1   Preliminary unit quantities

102

			I	QTY	±	DESCRIPTION CODE	
Preliminary unit quantities()							
	(ref. S Ø CB)		V		V	V	V
1	Flush door size 44 × 826 × 2040 with skeleton core and		W		V	V	V
			E	F	J	P	L
	plywood 6.s. and h/wd lipping long edges		I	QTY	±	DESCRIPTION CODE	
			V		V	V	V
			W		V	V	V
	&		E	F	J	P	L
	Pair 75 mm mild steel butts to softwood		I	QTY	±	DESCRIPTION CODE	
			V		V	V	V
			W		V	V	V
	&		E	F	J	P	L
	Rim lock (ref. RL 123) to softwood		I	QTY	±	DESCRIPTION CODE	
			V	·	V	V	V
			W		V	V	V
			E	F	J	P	L
	$2/65 \frac{826}{130}$ $\frac{956}{}$ $4\frac{2.040}{65}$ $\frac{2.105}{4.210}$ $\frac{956}{5.166}$ horns $\frac{150}{5.316}$		I	QTY	±	DESCRIPTION CODE	
			V		V	V	V
			W		V	V	V
			E	F	J	P	L
	100 × 75 softwood rebated and moulded door frame		I	QTY	±	DESCRIPTION CODE	
5.32			V		V	V	V
			W		V	V	V
			E	F	J	P	L
	Bed wood frame in cement mortar (1:3) and point one side		I	QTY	±	DESCRIPTION CODE	
5.17			V		V	V	V
			W		V	V	V
			E	F	J	P	L

Figure 9.2.2   Preliminary unit quantities

103

I	QTY	±	DESCRIPTION CODE	

<u>Preliminary unit quantities(3)</u>

(Ref. SØ CB)

I	QTY	±	DESCRIPTION CODE	
V		V	V	V
W		V	V	V
E	F	J	P	L

2/ 3

3 × 25 galvanised wrought iron fixing cramp 250mm girth once bent and twice holed

I	QTY	±	DESCRIPTION CODE	
V		V	V	V
W		V	V	V
E	F	J	P	L

2/ 0.83
2.04
0.04
5.74

KPS & (3) wood general surfaces (inty)

I	QTY	±	DESCRIPTION CODE	
V		V	V	V
W		V	V	V
E	F	J	P	L

5.17

Ditto wood frames 100-200mm girth (inty)

I	QTY	±	DESCRIPTION CODE	
V		V	V	V
W		V	V	V
E	F	J	P	L

I	QTY	±	DESCRIPTION CODE	
V		V	V	V
W		V	V	V
E	F	J	P	L

I	QTY	±	DESCRIPTION CODE	
V		V	V	V
W		V	V	V
E	F	J	P	L

I	QTY	±	DESCRIPTION CODE	
V		V	V	V
W		V	V	V
E	F	J	P	L

Figure 9.2.3   Preliminary unit quantities

each rogue description used must be included as a rogue whenever the unit quantity is used.

(e) Code the Items

Each item is coded in the normal way. These codes will be used for the unit quantity library to generate full descriptions. Great care must therefore be exercised at this stage since it is difficult to know what the library contains before a bill of quantities is produced.

(f) Prepare the Contents of the Unit Quantity Library

Items to be included in the unit quantity library are taken off by considering separately each item on the preliminary unit quantity dimension sheets. An individual item is needed for each valid combination of parameters that cause the item. As an example, figure 9.3 indicates the taking-off for items of brickwork necessary to cover the varying parameter combinations: S O CB, S C CB, D O CB and D C CB. Certain items may not be contained in the preliminary unit quantity dimensions. Although a combination of parameters may be chosen for the preliminary unit quantities that covers the widest possible range of items a different combination of parameters may introduce new items. For instance, the example chosen is based on a single unit; the introduction of a double unit creates a new item of brickwork for the partition dividing the stores.

(g) Check for Identical Items

Identical items created by different parameter sets need only be included once in the unit quantity library (appendix A). A check must be made for identical items taken off under the different parameter sets. Items common to parameter sets may be generated by reference only to the master code for that particular set of unit quantities, the parameter references being left blank.

A unit quantity reference consists of (i) a master code reference and (ii) parameter code references. The master code is unique for any set of unit quantities and identifies the component. The master code reference for the store units is assumed to be '542L'. Parameter code references should be kept as simple as possible and should be easily identified with the parameter sets. The notation used to identify the parameter sets therefore may be adopted as suitable parameter code references for the store units. Items common to a number of parameter sets need only be coded with the master code and those parameters which create the item. Thus, wherever parameter codes are indicated, the item is unique to the parameters referenced. For example, only two items are needed for the store doors in the unity quantity library (reference 542L) for the store units.

    (1) Item reference 542L S – ––; 1 No. Flush door (for a single store unit)

    (2) Item reference 542L D – ––; 2 No. Flush doors (for double store units)

			Final unit quanties (1)	I	QTY	±	DESCRIPTION CODE	
4.25 2.15	11.69		HBW in c.6. in ct. mo. (1:3) fin with ff. o.s. (S ∅ CB)	V	V	V	V	
				W	V	V	V	
				E	F	J	P	L
2.75 2.75	7.56	Ditto	1.500 1.250 2.750 (S C CB)	I	QTY	±	DESCRIPTION CODE	
				V	V	V	V	
				W	V	V	V	
				E	F	J	P	L
0.96 2.18	2.09	(door) Ddt last two items (S - CB)		I	QTY	±	DESCRIPTION CODE	
				V	V	V	V	
				W	V	V	V	
				E	F	J	P	L
5.60 2.75	15.40	2/1.500 3.000 2.603 5.603 HBW in c.6. in ct. mo. (1:3) fin with ff. o.s. (D ∅ CB)		I	QTY	±	DESCRIPTION CODE	
				V	V	V	V	
				W	V	V	V	
				E	F	J	P	L
4.10 2.75	11.28	Ditto	2.603 1.500 4.103 (D C CB)	I	QTY	±	DESCRIPTION CODE	
				V	V	V	V	
				W	V	V	V	
				E	F	J	P	L
2/ 0.96 2.18	4.19	(doors) Ddt last two items (D— CB)		I	QTY	±	DESCRIPTION CODE	
				V	V	V	V	
				W	V	V	V	
				E	F	J	P	L
1.40 2.75	3.85	HBW in c.6. in ct. mo. (1:3) fin ff. 6.s (D— CB)	1.500 103 1.397	I	QTY	±	DESCRIPTION CODE	
				V	V	V	V	
				W	V	V	V	
				E	F	J	P	L

Figure 9.3   Final unit quantities

The parameter references indicated by a (−) would be left blank in the unit quantity library. Item (1) therefore will be generated whenever a single unit is coded irrespective of parameters 2 and 3 for the following parameter sets: S O CB, S C CB, S O BL, and S C BL. Also, item (2) will be generated whenever double units are coded for parameter sets: D O CB, D C CB, D O BL, and D C BL. Similarly, the reference S C −− indicates that the item associated with this code is common to all situations in parameter 3 (that is, brick or concrete block) when in a single corner position; the items associated with the reference S − −− are common to all single units whether they are in an open or corner position (parameter 2) or whether constructed of brick or concrete block (parameter 3).

(h) Prepare a Guide-sheet

The quantities indicated in appendix A are stored by the computer and are therefore lost by the quantity surveyor except by reference to the final unit quantity dimension sheets. A *guide-sheet* must therefore be prepared for reference when taking off unit quantities. This sets out the details of the parameter coding as shown in the following schedule.

UNIT QUANTITY FOR STORE
MASTER CODE 542L

Parameters 1−3 in columns 34−37 inclusive

	Parameter			
	1	2	3	
Columns	34	35	36−37	Interpretation of characters
	S			Single unit
	D			Double unit
		O		Situated on an open wall
		C		Situated in a corner position
			CB	Constructed of common bricks
			BL	Constructed of concrete blocks

The library has been limited to the measurement of

(i) walls to form store units
(ii) plasterwork to walls
(iii) rendering to store internally (where blocks specified)
(iv) doors, frames and ironmongery
(v) decoration to woodwork and plasterwork

The library does not include

(i) floor and ceiling finishes
(ii) external hollow walls

The master code and full range of parameters should be set down clearly and precisely with sketches as an expedient method of explaining the parameters when found necessary. The work covered by the unit quantities should be described in sufficient detail to remove the necessity for reference to the unit quantity dimension sheets. Any work included in the unit quantity library but not directly associated with the component should also be described on the guide-sheet. Similarly any work that needs to be dealt with by ordinary taking-off must be described.

The quantity surveyor should be provided with a check-list of the items contained in the unit quantity library. This acts as an aid when taking-off to enable the unit quantities to be dovetailed into the measurement as a whole. The following example illustrates the way in which this may be achieved.

CHECK LIST OF UNIT QUANTITIES – REFERENCE 542L

The following items have been measured in connection with the standard store units.

(1) Lintels
(2) Brickwork or blockwork
(3) Bedding wood frames
(4) Flush doors
(5) Door frames
(6) Hinges and ironmongery
(7) Fixing cramps
(8) Plasterwork to external face of store
(9) Rendering to internal face of store (blocks only)
(10) Painting to doors and frames
(11) Decoration to plasterwork

The following items are not included in the unit quantities.

(1) Floor and ceiling finishes
(2) External cavity walls

CODING UNIT QUANTITIES DIMENSIONS

The dimensions of the final unit quantities for input to the unit quantities library file are coded in two parts. The first part (A) contains the description coding for the production of a full standard library description, which corresponds to the coding of normal dimensions. The second part (B) contains the master code and parameter coding for identification of unit quantities.

The dimension sheets have an extra column that provides coding boxes for recording the master code and parameter codes (column 6, figure 9.4). Thus two code cards are provided for (a) description codes card A and (b) unit quantities codes card B.

Card A is a dimension card that contains all the coded information relating to the item description and contains the normal dimension coding. A reference is also made to the fact that a link exists with the information contained on an annotation card (card B) and that it should be read in conjunction with that card (figure 9.4, line references 1A and 1B). The annotation card contains the information necessary to identify the parameters and their relevant unit quantities. The references on card B are as follows.

108

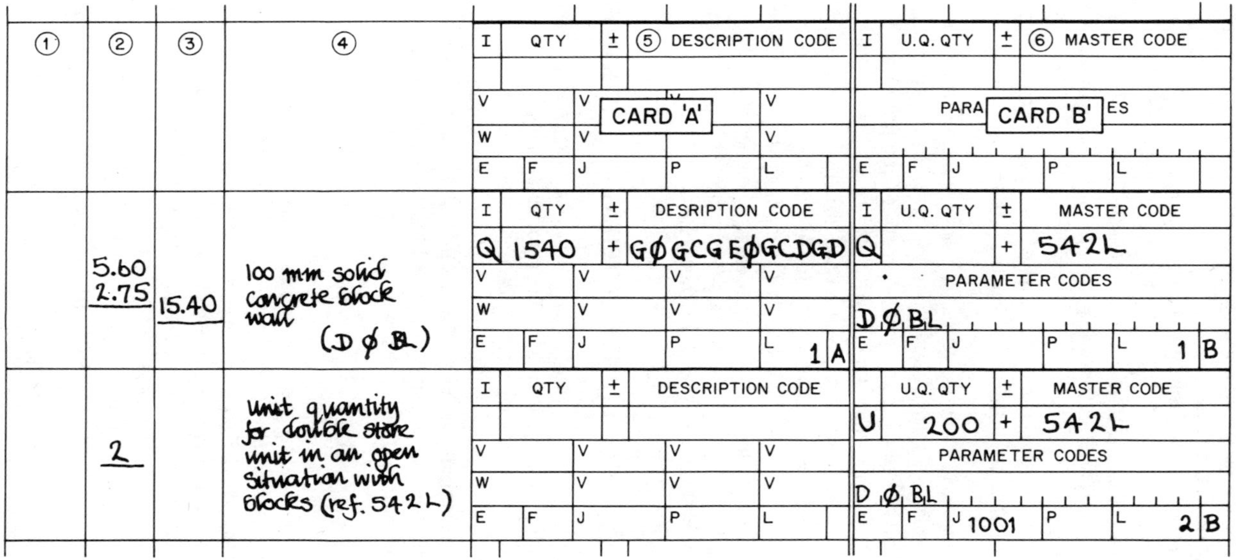

①	②	③	④	I	QTY	±	⑤ DESCRIPTION CODE	I	U.Q. QTY	±	⑥ MASTER CODE
				V		V	CARD 'A' V		PARA CARD 'B' ES		
				W		V	V				
				E	F	J	P L	E	F J	P	L

	5.60 2.75	15.40	100 mm solid concrete block wall (DØBL)	I Q	QTY 1540	± +	DESRIPTION CODE GØGCGEØGCDGD	I Q	U.Q. QTY	± +	MASTER CODE 542L
				V		V	V		PARAMETER CODES		
				W		V	V	DØBL			
				E	F	J	P L 1 A	E	F J	P	L 1 B

	2		unit quantity for double store unit in an open situation with blocks (ref. 542L)	I	QTY	±	DESCRIPTION CODE	U	U.Q. QTY 200	± +	MASTER CODE 542L
				V		V	V		PARAMETER CODES		
				W		V	V	DØBL			
				E	F	J	P L	E	F J 1001	P	L 2 B

Figure 9.4   Coding unit quantities

(i) A reference that the input is for the unit quantity library.
(ii) An indication of the type of entry on to the unit quantity library. The item may be required to (1) be deleted from the library, (2) be added to the library, (3) change an existing entry, or (4) delete all the entries under the master code indicated.
(iii) The job reference and any bill element or feature references (code fields J, E and F).
(iv) The master code for the component (542L).
(v) The unique parameter codes required to call up the items (see parameter guide-sheet).
(vi) The page and line references indicating the sequence of the item; this should correspond with card A for that item.

Details of unit quantities codes on cards A and B are shown in figure 9.4. The quantities as coded on cards A and B are input to the computer and are recorded on a file-holding device by a program written for the purpose.

The main difficulty when dealing with unit quantities lies in coding the units. The quantity surveyor must be given all the information necessary for the description of a component to enable him to select the right parameters. It is therefore good practice for the taker-off to code unit quantities as he takes them off. In this way he is assured that the correct references have been made. There is great difficulty in tracing errors of coding when editing a unit quantities bill since they may be easily masked by other items.

BILLS OF QUANTITIES FROM UNIT QUANTITIES

A bill of quantities may be produced from the unit quantity library by using card B with references to the unit quantity master code and the unique parameter codes. The unit of measurement is 'number' and the unit quantities are multiplied by the number of components involved when more than one component is required. For example, assume that quantities are required for two (2 NO) double store units constructed of blockwork in an open position. The code references inserted on card B would be as follows.

(a) A reference to the type of input indicating a unit quantity; assume this to be (U).
(b) The number of components involved; this is usually expressed as a whole number. However, the quantity field is usually expressed to two decimal places in the event that a fraction of a component is required. The quantity involved therefore is expressed to two decimal places (2.00).
(c) The arithmetic value of the measurement is addition or (+).
(d) The unit quantity master code (542L).
(e) The unique parameter code reference (D O BL).
(f) References relating to location and line numbers are dealt with in the same manner as normal dimensions.

Card B therefore would be filled in as follows: U 200 + 542L D O BL (figure 9.4). The computer will identify the input as unit quantities by the first reference U and locate the items in the unit quantity library by using the master code 542L. Each item is scanned and the unique parameter codes examined and

110

checked with the references on card B (D O BL). Any item with a comparable code is extracted and its quantity multiplied by 2.00 and recorded for output.

On examination of the items contained in appendix A it will be apparent that the first item to be selected is item 2.

  (i) 2.32 M   'Precast concrete; normal; mix 1:2:4 20 mm aggregate; lintels; bedding in cement mortar (1:3); 102 x 75; reinforced one 10 mm bar'   (Ref. D ———)

The quantity of this item is then increased to 4.64 M (2.32 x 2) and rounded off to 5 M for output.

The second item to be selected is item 10.

  (ii) 15.06 M2   'Solid concrete blocks; 102 mm thick; walls or partitions'
                                      (Ref. D O BL)

This is increased to 30.12 M2 and rounded off to 30 M2 for output.

The third item to be selected is item 14.

  (iii) 8.25 M   'Bonding ends to brickwork; 102 mm blockwork'
                                      (Ref. D – BL)

This is increased to 16.50 M and rounded off to 17 M for output.

This process is continued until all the items have been examined and the relevant items extracted, their quantities multiplied by 2 and recorded on an output file for print-out. The final print-out will therefore contain the following quantities.

Unit Quantities for 2 No. Stores (Reference 542L D O BL)

Quantity	Description
5 M	Precast concrete; normal; mix 1:2:4 20 mm aggregate; lintels; bedding in cement mortar (1:3); 102 x 75; reinforced one 10 mm bar
30 M2	Solid concrete blocks; 102 mm thick walls or partitions
17 M	Bonding ends to brickwork; 102 mm blockwork
14 M	100 mm blockwork against soffits
21 M	Bedding in cement mortar (1:3); wood frames or sills
4 NO	Flush door; skeleton core; 44 x 826 x 2040 mm; plywood facing both sides; hardwood lipping long edges
21 M	Softwood; door frame; 100 x 75 mm; rebated and moulded labours
4 PR	To softwood; mild steel; butts; 75 mm
4 NO	To softwood; rim lock; ref. RL123
24 NO	To softwood; wrought iron; galvanised; cramps 3 x 25 x 250 mm girth; bent once; holed twice
50 M2	Mortar; cement and sand (1:3); steel trowelled; to walls; over 300 mm wide
28 M2	Plaster; first coat gypsum plaster undercoat and sand; finishing coat gypsum plaster finishing steel trowelled; to walls over 300 mm wide
15 M2	Painting one coat lead free wood primer; two undercoats, one coat alkyd based paint full gloss finish; wood surfaces; general surfaces; over 300 mm girth
21 M	Painting one coat lead free wood primer; two undercoats, one coat alkyd based paint full gloss finish; frames or the like; over 100 mm not exceeding 200 mm girth
28 M2	Painting three coats emulsion wall finish to plaster

Unit Quantity Variables

The creation of a unit quantities library, like the standard library of descriptions, presents a problem of size. A reference must be made of each item under every possible situation and variations in the specification or measurement of a single unit quantity item may occur, thus increasing the size of the library. This problem may be overcome if the library contains parameters with skeleton quantities that can be sorted by the computer to produce total unit quantities for a component in any particular situation.

The unit quantities library may be compiled to take advantage of certain consistent measurement and description patterns that emerge within individual sets of unit quantities. For instance, identical items may occur in several unit quantities within any one range of components. In the example illustrated (unit quantities for store units, appendix A) the doors are a constant commodity and required under all parameter situations. These items need no parameters to generate them and appear in every version of the unit quantities. In the example they need only be included once for all parameters relating to single or double units (see items 24 and 25 appendix A).

Description Variables

Additional facilities have been written into programs to-provide routines that will generate individual unit quantities from the skeleton unit quantities on file. Identical items may occur, which vary according to the choice of the types or sizes of materials used; for instance the type of plaster may vary as may the specification of paintwork. Programs have been written to generate full descriptions from parameters used as part of the description code. This facility is used where there is a large choice of materials and the necessity of having to enter each different type of material in every possible situation is thus avoided. These variables appear in the description of the unit quantities items in the library and may be inferred by a character or characters such as asterisks (**), which the computer will recognise as a variable insert. The following items may therefore appear in lieu of the items in the unit quantities library where the choice of materials is a variable.

(i) In lieu of items 36 to 39 (appendix A)
'Plaster; **; to walls; over 300 mm wide
(ii) In lieu of items 40 to 43 (appendix A)
'Painting; ***; general surfaces; over 300 mm girth
'Painting; ***; frames or the like; over 100 mm not exceeding 200 mm girth

The parameter reference to replace the asterisks should be the code contained in the standard library of descriptions. The selection of the variable is made by coding card B with the appropriate variable parameter code.

In the same way the sizes of materials may vary and a parameter may be introduced in the description code to infer the sizes. For example, the doors for the store units may vary in size. Thus the following item may appear in lieu of the items in the unit quantities library where a choice of the sizes of doors is required.

(iii) In lieu of items 24 and 25 (appendix A)
'Flush door; skeleton core; 44 x * x **; plywood facing both sides; hardwood lipping long edges

Dimensions Variables

Identical items may have the same descriptions but may vary in their quantity. For example a parameter may be introduced to vary the floor-to-ceiling height of the store units. The optimum coverage of items possessing this characteristic may be achieved by implying the parameter variable in the dimension field of the unit quantities. The item in the unit quantities library may therefore appear as follows.

**	4.25 M2	Selected common bricks; in cement mortar (1:3); half brick thick; flush pointing one side; walls; stretcher bond	542L S O CB
**	2.75 M2	Selected common bricks................... stretcher bond	542L S C CB
**	5.60 M2	Selected common bricks.................. stretcher bond	542L D O CB
**	4.10 M2	Selected common bricks.................. stretcher bond	542L D C CB

The asterisks represent the variable parameter for the floor-to-ceiling heights. The dimensions represent the mean girths of the walls.

Parameter Look-up Table

A parameter look-up table is necessary to help the computer pick up the correct variables. This means that the variables must be coded for easy reference. For example, assume that the floor-to-ceiling heights in parameter 2 for the store units vary in 0.25 m stages commencing at 2.50 m up to a maximum height of 3.75 m. The following information would be fed into the computer as a parameter reference for the variable heights.

 (i) A reference that the input is a parameter look-up table; assume this reference to be P.
 (ii) An indication of the type of entry on to the unit quantities library; assume the reference for an addition to the library is A.
(iii) A reference to the parameter to which the table relates; this is the second parameter and therefore will be 2.
(iv) The master code for the component is 542L for the store units.
 (v) The variable equivalents or code references; these are assumed to be A to F.
(vi) The parameter variables are the floor-to-ceiling heights.

Assuming the foregoing information the look-up table would be set up by entering a batch of data as follows.

113

(i)	(ii)	(iii)	(iv)	(v)	(vi)
P	A	2	542L	A	2.50
P	A	2	542L	B	2.75
P	A	2	542L	C	3.00
P	A	2	542L	D	3.25
P	A	2	542L	E	3.50
P	A	2	542L	F	3.75

Unit quantities may then be generated by using the above variable equivalents (column v) as the code reference to replace the asterisks.

Use of Unit Quantities

The use of unit quantities is one of the advantages gained by the introduction of standardisation through data coordination, which is outlined in chapters 6 and 7. This facility is particularly applicable to methods of system building or industrialised building made up of standard components. However, unit quantities may be used wherever standard details are available for the construction of buildings. The system is made possible only by the use of the computer and standardisation of details.

The use of the computer is an important development in the process of data preparation. If it were not for the facilities that the computer can provide the application of unit quantities would be difficult. The computer's power of storage and its ability to retrieve information rapidly are features that the quantity surveyor finds of great value.

Standardisation of details involves the preparation of standard designs for components and the use of standard specifications. The use of unit quantities not only helps the quantity surveyor to prepare his bills of quantities but also helps the architect to prepare his details. No longer need the architect produce schedules for use by the quantity surveyor. These may be replaced by a standard code, which can be extracted from the drawings and transferred directly to the parameters of the quantity surveyor's unit quantities.

REFERENCES

1. —— *Computer System Handbook*, Part 3M, Consortium of Local Authorities Special Programme: General Development Group.
2. —— *Computer System General Introduction*, Local Authorities Management Services and Computer Committee: Steering and Development Group.
3. K. W. Monk and T. D. Dunstone, 'Two Years of Practical Quantity Surveying by Computer', *Chart. Surv.*, 98 (1966) pp. 363–8.
4. —— *Computer Techniques Billing Systems* (R.I.C.S., London, 1973).

10 THE USE OF THE COMPUTER FOR ANCILLARY SERVICES

The development of computer techniques for quantity surveying services was initially concerned with systems for the production of bills of quantities. With the successful application of computerised systems to bill production the quantity surveyor has turned his attention to possible applications in other areas of quantity surveying.

BILLS OF QUANTITIES

The bill of quantities is a starting point for the quantity surveyor's analytical activities. By adopting computer techniques the measured data can be reprocessed to provide information that will be beneficial to the design team and the contractor. The amount of work that is performed by the computer varies according to the extent of the particulars provided at the measuring stage. Better results may be achieved by extending the systems developed for bill production or by providing the information as part of a computerised coordinated information system. Bill-production programs may be modified to produce results to suit specific requirements of a project. For instance, the whole or part of the measured work can be adjusted by introducing a timesing factor and output may be modified to produce combined documents. A specification may also be produced by applying a routine that suppresses the quantities relating to each measured item, while a schedule of rates may be produced by adding prices to the items in the specification.

The computer may be used to speed production and to lower costs in the provision of the following services.

(a) Bills of quantities sortation
(b) Bill pricing
(c) Cost planning and cost analysis
(d) Network and construction analyses involving financial and resource analysis for design and operational control, materials ordering, labour and plant allocation and bonusing.
(e) Financial forecasts, interim valuations and final accounts.

The systems used are many and varied and it is only possible at this stage to study the main characteristics and benefits of some of the systems in use.

BILL SORTATION

Bills of quantities sortation is one of the first considerations of the quantity surveyor when planning his analytical activities. The formats of the bills can conveniently be sorted from the original measured work to satisfy the requirements of the quantity surveyor's ancillary services and the needs of the contractor. Bills of quantities are now being used for cost planning, thus providing a service to the design team. Use is also being made of the bills by the contractor as a managment tool. For instance, a bill of quantities in a *trade-by-trade* or *sectional* order may be used for tendering purposes, an *elemental* format may be used for cost planning, and an *operational* format may be used for the contractor's analytical activities. Its presentation in an elemental or operational format forms a basis for management control and affords a breakdown of labour and materials for use when programming the work and for cost accounting and bonusing. In this respect the computer may assist in implementing a coordinated information system. The routines applied to enable the computer to provide a suitable sortation generally form part of the programs used for bill production.

The prospect of changing the bills of quantities into a foreign language may be a matter of vital importance with the United Kingdom's connections with the European Economic Community. This is possibly where the computer would be able to provide an invaluable service and such a facility should be developed. The system could conceivably be reduced to a procedure merely involving the change of library tapes.

BILL PRICING

A priced bill of quantities may be produced by feeding the computer with bill references and appropriate unit rates. The unit rates may be entered by using one of two methods.[1] The routines applied may enter the unit rates on to the job file via the dimension sheets or alternatively by means of an updating routine. Both these methods can be used on the same project. For instance, the rates that are input via the dimension sheets will produce an estimate and for contract purposes the job file may be amended by inserting the contractor's rates by using an updating routine. It is also possible to carry rates on to the short code and unit quantities libraries wherever this is deemed to be appropriate.

The bill-pricing facility enables an estimate to be prepared for the purpose of tender evaluation. The tender figure may then be checked for its validity and reasonableness of price. Prices used for tendering may be inserted against the items stored on the job file. The computer can then check the contractor's extensions or a bill rate listing may be produced, which enables the quantity surveyor to check the rates and extensions against the original tender. A priced and extended bill of quantities may be produced subsequent to these routines. Programs are also available that provide a comparison of rates between tenders and if the need arises comparisons can be made between the various sections of a tender.[2] Any inconsistencies and differentials in rates will be highlighted.

COST PLANNING AND COST ANALYSIS

Cost planning is a method of controlling the cost of a building within a predetermined value during the design stage. This involves the preparation of an approximate estimate to determine the cost limit for a building and the systematic breakdown of this figure into elemental costs to formulate a cost plan. The cost plan for a project is a document that states the intended expenditure for each element for design purposes in relation to a defined standard of quality. The details contained in the approximate estimate and cost plan are usually compiled from records of costs produced by cost analyses of previous projects. A cost analysis is the systematic breakdown of cost data into recognisable divisions or elements of a building based on the costs contained in the priced bills of quantities for a project.

The computer's facilities enable cost analysis programs to be used for presenting cost plans and cost estimates, which may be produced in a form that promotes cross-referencing. One of the benefits to be gained by using the computer is its ability to produce multiple analyses quickly and economically from raw data. Different projects within the same building type may also be compared in terms of both their building economy and planned efficiency. An example of a typical computer print-out generated from a cost analysis program

Table 10.1   Cost analysis print-out

COST ANALYSIS :	BLOCK/ELEMENT			CONTRACT 046
ELEMENT		TOTAL	PERCENT	COST/M2
1 PRELIMINARIES		10840	10.93	8.14
2 CONTINGENCIES		3000	3.03	2.25
3 SUBSTRUCTURE		6751	6.81	5.07
4 FRAME		9081	9.16	6.82
5 UPPER FLOORS		5395	5.44	4.05
6 ROOFS		3757	3.79	2.82
7 STAIRCASES		1951	1.97	1.46
8 EXTERNAL WALLS		6365	6.42	4.78
9 WINDOWS AND EXTERNAL DOORS		4175	4.21	3.13
10 INTERNAL WALLS AND PARTITIONS		2913	2.94	2.19
11 INTERNAL DOORS		2260	2.28	1.70
12 WALL FINISHES		2650	2.67	1.99
13 FLOOR FINISHES		4016	4.05	3.02
14 CEILING FINISHES		3995	4.03	3.00
15 FITTINGS AND FURNISHINGS		1250	1.26	0.94
16 SANITARY APPLIANCES		719	0.73	0.54
17 DISPOSAL INSTALLATIONS		628	0.63	0.47
18 WATER INSTALLATIONS		1586	1.60	1.19
19 HEAT SOURCE		6295	6.35	4.73
20 VENTILATING SYSTEM		385	0.39	0.29
21 ELECTRICAL SERVICES		6642	6.70	4.98
22 GAS SERVICES		325	0.33	0.24
23 LIFT AND CONVEYOR INSTALLATION		6400	6.46	4.80
24 BUILDERS WORK IN CONNECTION WITH SERVICES		1461	1.47	1.10
25 BUILDERS PROFIT IN CONNECTION WITH SERVICES		162	0.16	0.12
26 SITE WORKS		261	0.26	0.20
27 DRAINAGE		895	0.90	0.67
28 EXTERNAL SERVICES		610	0.62	0.46
29 MINOR BUILDING WORKS		4375	4.41	3.28
		------	------	-----
GROSS AREA 1332 M2 TOTAL		99143	100.00	74.43

Table 10.2   Block/element analysis print-out from a cost analysis program

COST ANALYSIS		ELEMENT/ITEM		CONTRACT 046
OFFICE BLOCK				
ELEMENT 8		EXTERNAL WALLS		
ITEM NO	QUANTITY	UNIT	RATE	TOTAL
27A	358	M2	6.46	2312.68
27B	134	M2	8.32	1114.88
27C	84	M2	7.26	609.84
31J	369	M	5.50	2029.50
40D	40	M2	7.45	298.00

				6364.90
			PERCENTAGE	6.42%
GROSS AREA	1332 M2	COST PER SQUARE METRE		£4.78
ELEMENT AREA	616 M2	COST PER SQUARE METRE		£10.33

is shown in table 10.1.[3] This lists the elements used and indicates (a) the value of each element, (b) the value of each element expressed as a percentage of the total cost and (c) the cost of each element per square metre of floor area. The list of elements is usually left to the discretion of the user.

Although the block/element analysis may be produced manually, the computer may be used to provide added information to obtain more detailed element/item analyses. These list the items within each element indicating the quantity, rate and total value of each item (table 10.2). Each item forming element 8 is shown individually. The items are identified by reference to the bill item numbers and indicate the total quantity of each item, its rate and total value. A summary of the costs of the element per square metre relative to the gross area of the building and the element area are also indicated. This facility enables the elemental analysis to be related more directly to the source documents by reference to each individual item. The information gained by the cost analysis exercise may be summarised in a print-out that produces over-all details of the various aspects of the project (table 10.3). This information is useful as a feedback and enables a potential user to decide on the relevance of the information in order to make intelligent use of it.

Cost planning by computer is usually based on an elemental analysis produced as a by-product when processing the bills of quantities using rates supplied from previous projects. The techniques vary and the computer can only be used if the input data carry sufficient details and references to enable the dissemination of the data into the relevant parts. The cost planning references should permit the relevant bill sortation from the records on the job file for the project. For this purpose the quantity surveyor will usually annotate his dimensions to enable the relevant information to be collated correctly.

The computer is able to compile a planned contract expenditure or predicted cash flow for a building project[4] (table 10.4). It is possible to use the computer

118

Table 10.3   *Summary of a cost analysis program*

```
COST ANALYSIS : SUMMARY                        CONTRACT 046

TENDER AMOUNT  £99143                          DATE  15 10 75

FINAL ACCOUNT                                  DATE

PRELIMINARIES    £10840        8.14%

TENDER ADJUSTMENT         NIL

CONTRACT PERIOD     60 WEEKS

ANALYSIS AREA      NET INTERNAL

BLOCK                          AREA M2            VALUE

00 SITE WORKS                                     261

01 MAIN OFFICE BLOCK            1332             94507

03 GARAGES                       90              4375

REMARKS

CONSTRUCTION.      STRUCTURE : STRUCTURAL STEEL WITH CONCRETE CASING

             FLOORS & ROOF : CONCRETE ON METAL DECK

                  CLADDING : BRICK INFILL PANELS

                  FINISHES : PLASTERBOARD DRY LINING

                   HEATING : LOW PRESSURE GAS FIRED

                  LIGHTING : EMERGENCY LIGHTING AND FIRE ALARM SYSTEM

          SPECIAL SERVICES : ONE 10 PERSON LIFT SINGLE SLIDING DOOR CAR ENTRANCE

SITE.  1 : 10 AVERAGE GRADIENT WITH ROCK OUTCROP BELOW MAIN BLOCK.   POOR ACCESS

     WITH RESTRICTION ON WORKING SPACE AND STORAGE OF MATERIALS

TENDER.  LOWEST OF EIGHT BY 0.18% BELOW SECOND AND 0.21% BELOW THIRD

CONTRACT DOCUMENTS.  DRAWING NOS  046/1 TO 046/16

                  BILLS OF QUANTITIES
```

to simulate various cost analyses resulting from different schedules. A project schedule of cost may be developed to take into account any resource and time limitations that may be specified. Cost optimisation may be achieved by using the computer to analyse resources. This involves a more specific approach to cost analysis. For instance, the contract period for a project may be shortened thereby creating a need to intensify the resources. The project may have to be completed by applying additional resources, which will mean a corresponding increase in the cost of all the activities affected.

Table 10.4 is a print-out of a typical cost analysis program showing the planned expenditure for a project in the form of a histogram.[5] The particulars indicate

(i) the period or day shown in column DATE
(ii) the planned expenditure for the period or day shown in column P
(iii) the planned cumulative expenditure for the period or day shown in column C
(iv) the number of weeks/days from the base data (14 November 1974) shown in column TIME

Table 10.4   Planned expenditure print-out

COST OUTPUT (PLANNED)   RUN 1   TIME NOW   14NOV74   PAGE 1

```
                              0      10000    20000    30000    40000    50000

DATE        P       C         I.........I.........I.........I.........I.........I      TIME

 7JUL75    980      980       IP        I         I         I         I         I      .1
 8JUL75    980     1960       I P       I         I         I         I         I      .2
 9JUL75    980     2940       I  P      I         I         I         I         I      .3
10JUL75    980     3920       I   P     I         I         I         I         I      .4
11JUL75    980     5150       I    P    I         I         I         I         I      .5
14JUL75   1210     6110       I     P   I         I         I         I         I      1.0
15JUL75   1210     7320       I      P  I         I         I         I         I      1.1
16JUL75   1210     8530       I       PPI         I         I         I         I      1.2
17JUL75   1210     9740       I       PI          I         I         I         I      1.3
18JUL75   1210    10950       I      PP           I         I         I         I      1.4
21JUL75   1340    12290       I        I P        I         I         I         I      2.0
22JUL75   1340    13630       I        I PP       I         I         I         I      2.1
23JUL75   1340    14970       I        I  P       I         I         I         I      2.2
24JUL75   1340    16310       I        I   P I    I         I         I         I      2.3
25JUL75   1340    17650       I        I    PP I  I         I         I         I      2.4
28JUL75   1180    18830       I        I      PI  I         I         I         I      3.0
29JUL75   1180    20010       I        I         P         I         I         I      3.1
30JUL75   1180    21190       I        I         IP        I         I         I      3.2
31JUL75   1180    22370       I        I         I P       I         I         I      3.3
 1AUG75   1180    23550       I        I         I PP      I         I         I      3.4
 4AUG75   1210    24760       I        I         I   P I   I         I         I      4.0
 5AUG75   1210    25970       I        I         I    P I  I         I         I      4.1
 6AUG75   1210    27180       I        I         I     P I I         I         I      4.2
 7AUG75   1210    28390       I        I         I      P I         I         I      4.3
 8AUG75   1210    29600       I        I         I       PI I         I         I      4.4
11AUG75   1180    30780       I        I         I         PP        I         I      5.0
12AUG75   1180    31960       I        I         I          I P      I         I      5.1
```

P : PLANNED COST   TIME : WEEKS/DAYS
C : CUMULATIVE PLANNED COST

NETWORK AND CONSTRUCTION ANALYSES

Network is a general term used to cover all those techniques and methods that depict a project by means of an arrow diagram showing the sequence and interrelationship of activities and events. It is used in the programming and control of construction projects. The technique enables the design activities to be logically programmed and effectively controlled. It involves the dissemination of the project into component activities from which their interrelationship and logical sequence of operation can be established. The techniques may be applied by using either the Critical Path Method (CPM) or the Programme Evaluation and and Review Technique (PERT).

The application of the foregoing techniques provides analyses of time, resources and cost. One of the major advantages of using such techniques is that problems of logical sequence can be separated from those involving time and resource allocation. In this way the resources available for the project can be deployed in the best possible manner. The critical activities that determine the project duration are also highlighted and provide a valuable contribution to the control of the project at the programming and construction stages.

Use may be made of the CPM/PERT methods of planning and scheduling by the contractor and the design team. The method has been adopted by many contractors who have utilised the more sophisticated facilities. These are wide-ranging and vary from the checking of monthly valuations, via the cost reporting aspect of the system, to the substantiation of claims for extension of time. A long-range program can be provided in greater detail and with the updating facility a clearer definition of problem areas can be identified at an early stage.

The involvement of the design team in CPM/PERT scheduling can be as much or as little as is desired. The design team often requires the contractor to produce schedules indicating the project programme and progress. The system may also be used in scheduling the programme of activities when planning the project.

The procedures to be followed when applying the techniques are listed below.

(a) Prepare a list of activities.
(b) Draw an arrow diagram to represent the sequence of activities.
(c) Estimate the duration of time for each activity.
(d) Analyse the arrow diagram and calculate
   (i) the earliest and latest permissible times for the performance of each activity
  (ii) the minimum project time and establish the sequence of activities that are critical in the performance of the project to provide the most expedient method of execution
 (iii) the amount of time by which an activity may be delayed or extended. This is known as *float-time* or *slack* and is associated with the non-critical activities.
(e) Prepare schedules listing the order in which each activity is to be performed.
(f) Allocate resources and prepare suitable schedules. It is from the foregoing information that an allocation of resources is made. This means that the duration of each of the various activities is adjusted within the float-times so that excessive demands on manpower and other resources are manipulated to achieve an optimum utilisation.
(g) Review the work in progress and update the information. The work can be controlled as it proceeds by using the network analyses to revise the work programmes to meet changing circumstances. A regular review and continual adjustment of the programme is needed to ensure that attention is directed at the activities to maintain effective control.

Programs are available to perform procedures d, e, f, and g and the computer can be used to analyse the arrow diagrams to calculate the timing of the activities and to prepare schedules (figure 10.1). However, for some projects the computer may be considered unnecessary and the network analysis may be carried out manually. It is on large and complex projects that computers are considered to provide particular advantages. Computer programs are designed to handle large quantities of data or to make long and involved calculations that produce a large number of results from a small amount of input. An analysis of an arrow diagram and the establishment of a critical path involves a combination of these two facilities.

The data are taken from an arrow diagram but before they are input the network must be checked for any duplication of references or the presence of any repetition of or break in the sequence of activities. Most programs contain an error-checking routine that will bring to light any such errors. However, it is possible to check the network manually if the number of activities is small.

The computer print-outs may consist of numeric schedules or bar charts, which deal with time analyses, resource analyses and cost analyses of the project. A choice of format is usually made available whereby the user can select the particular type of print-out he wants to suit his needs. Each of these may vary in a number of ways, from those showing key points in the project for use by top management, to those showing more detail for use by the supervisory staff.

A comprehensive set of linked computer programs has been produced by International Computers Ltd (ICL)[4] for use on project planning and control The benefits that are gained by using these programs may be considered under the following headings.

(1) Time analysis
(2) Resource allocation
(3) Cost control

Time Analysis

The print-outs produced by the time-analysis program enable management to plan projects on a time basis. Standard print-outs may be obtained to show activity sequences in a number of different ways. For example, a standard schedule may be produced which shows an analysis of the activities indicating a sequence of activities with total float- and earliest start-times (table 10.5). This shows the total float-time as the major sequence of activities. Each activity is listed in the order of the amount of float time available from zero float-time upwards. The minimum project duration and the critical path may be established from this information. The critical path defines the sequence of activities that are critical in the performance of the project. These are found by examining the float time and are indicated by those activities with a zero float.

Another print-out may be obtained with the *report code* or *subproject* as the major sequence. This means that the schedule can be split into sections with each report code or subproject as the main division. The sequence of activities in each section is in order of the earliest start-time. A new page is produced for each section, which means that each person responsible for that section of the project may have a copy of the schedule relating to that part of the project for which he is responsible.

A bar chart may also be produced indicating a time analysis in report code and earliest start sequence (table 10.6). This is a convenient form of report that shows the result of a time analysis. It is also useful as a working document in monitoring the progress of a project. Bar charts may be produced in stages where frequent updating runs are applied. These show the progress of a project covering the total duration between each updating run. This means that any changes that occur in the planned schedule are incorporated at each stage.

A simple example of a time analysis is taken from figure 10.1 which shows an arrow diagram indicating typical activities relating to the development of a building project. The activities are indicated by arrows and the completion of an activity, which is known as an *event*, is indicated by a circle. The following are the input data for the analysis.

PRECEDING EVENT	SUCCEEDING EVENT	ACTIVITY	DURATION IN DAYS
A	1	Sketch plans and estimates	28
1	3	Consideration of plans	28
1	2	Outline town planning	21
2	3	Dummy	0
3	4	Completion of statutory approval	28
3	5	Subcontract details and estimates	42
3	6	Consultant's drawings	56
3	7	Working drawings	98
4	7	Dummy	0
5	7	Dummy	0
6	7	Dummy	0
7	8	Prepare bills of quantities	35
8	9	Print bills of quantities	7
9	10	Tendering	28
9	11	Annotate bills	7
11	12	Examine priced bills	7
12	13	Dummy	0
10	13	Consider tenders	21
13	14	Prepare contract documents	7
14	15	Sign contract	1

A dummy is a notional activity indicating a sequential relationship. It is a means of expressing the correct logic to illustrate the events that must be completed before proceeding. It is usually drawn with dotted or hatched lines for easy identification.

The computer processes the data by examining a sequence of event numbers and the duration of each activity from which a schedule is compiled of the earliest start/finish and latest start/finish times of each activity. A sort routine is then applied, which arranges the activities into sequences that may be specified by the user. The information contained on the output documents presents analyses of the data in varying sequences to suit the needs of the user. The following are typical sequential arrangements that may be obtained.

(a) In order of the event numbers; this sets out the details of the start/finish times in order of the events which can be read easily with the arrow diagram.

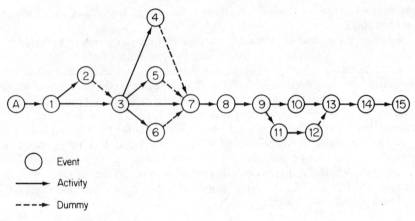

Event

Activity

Dummy

Figure 10.1   Network analysis — arrow diagram

(b) Earliest start-times (table 10.5); the print-out lists the activities and
    indicates the preceding event (P/E), succeeding event (S/E) and the
    duration of the activity (DUR) in days. Any float-time (FL) is also shown.
    The critical activities are found to be those with a zero float-time. The
    sequence is useful for checking the progress of the work with the
    programme.
(c) In order of float time; this emphasises the critical and near critical activities
    on which effort must be concentrated.
(d) By responsibility; this sets out the activities within areas of responsibility in
    order of the earliest start-times. The progress of the work may be checked
    against the programme within each area of responsibility.
(e) Bar chart (table 10.6); a graphical representation of the foregoing may be
    produced. The print-out shows the activities in the form of a bar chart which
    indicates the time factor by a chain of symbols as follows
    (i) critical activities (CCCCC)
    (ii) non-critical activities (AAAA)
    (iii) float-time (. . . . . . .)

Resource Allocation

Resource allocation is the process of adjusting the scheduled times and durations
of activities to suit the demands on resources or to keep these demands in step
with available resources. It is an empirical process involving the optimum
arrangement of each activity within the over-all pattern of the resources available
Difficulties do occur with the great amount of array required to provide an
acceptable solution. The main problem usually lies in the variety of resources
and the number of possible solutions. The listing of possible solutions becomes a
mechanical process that can best be performed by the computer, which permits
a more sophisticated approach to the problem. Projects may be subject to
limited resources or time restrictions. The computer will, in such events, smooth

124

Table 10.5 Time analysis print-out in earliest start sequence

PROJECT PLANNING

TIME ANALYSIS IN EARLIEST START SEQUENCE

TIME NOW   14NOV74

P/E	S/E	DESCRIPTION	DUR	EARLIEST START	EARLIEST FINISH	LATEST START	LATEST FINISH	FL
A	1	SKETCH PLANS & ESTIMATES	28	14NOV74	22NOV74	14NOV74	22NOV74	0
1	2	OUTLINE TOWN PLANNING	21	22DEC74	20JAN75	2JAN75	31JAN75	7
1	3	CONSIDERATION OF PLANS	28	22DEC74	31JAN75	22DEC74	31JAN75	0
2	3	DUMMY	0	20JAN75	20JAN75	31JAN75	31JAN75	7
3	4	COMPLETION STATUTORY APPROVAL	28	31JAN75	10MAR75	9MAY75	16JUN75	70
3	5	SUBCONTRACT DETAILS/ESTIMATES	42	31JAN75	30MAR75	19APR75	16JUN75	56
3	6	CONSULTANTS DRAWINGS	56	31JAN75	19APR75	30MAR75	16JUN75	42
3	7	WORKING DRAWINGS	98	31JAN75	16JUN75	31JAN75	16JUN75	0
4	7	DUMMY	0	10MAR75	10MAR75	16JUN75	16JUN75	70
5	7	DUMMY	0	30MAR75	30MAR75	16JUN75	16JUN75	56
6	7	DUMMY	0	19APR75	19APR75	16JUN75	16JUN75	42
7	8	PREPARE BILLS OF QUANTITIES	35	16JUN75	4AUG75	16JUN75	4AUG75	0
8	9	PRINT BILLS OF QUANTITIES	7	4AUG75	15AUG75	4AUG75	15AUG75	0
9	10	TENDERING	28	15AUG75	22SEP75	15AUG75	22SEP75	0
9	11	ANNOTATE BILLS OF QUANTITIES	7	15AUG75	24AUG75	30CT75	120CT75	35
11	12	EXAMINE PRICED BILLS	7	24AUG75	4SEP75	120CT75	230CT75	35
12	13	DUMMY	0	4SEP75	4SEP75	230CT75	230CT75	35
10	13	CONSIDER TENDERS	21	22SEP75	230CT75	22SEP75	230CT75	0
13	14	PREPARE CONTRACT DOCUMENTS	7	230CT75	1NOV75	230CT75	1NOV75	0
14	15	SIGN CONTRACT	1	1NOV75	2NOV75	1NOV75	2NOV75	0

P/E : PRECEDING EVENT
S/E : SUCCEEDING EVENT
DUR : DURATION OF ACTIVITY
FL : FLOAT TIME

out the resource allocation by making the necessary adjustments. For example, a project with a limited amount of manpower may have its duration extended to permit completion. Conversely the project may have an imposed time limit in which case the manpower and other resources will be increased.

Time analysis does not take account of any restrictions arising from the shortage of resources. It may therefore be undesirable to operate a project based only on the time-analysis schedules produced. The computer will apply routines that select alternatives and take account of any constraints imposed during the time span of the project. A resource-analysis program will produce schedules, histograms and bar charts indicating the various types of resource involving

125

Table 10.6   Bar chart print-out – time analysis

```
BAR CHART
                                    14        28        12        26        9         23
P/E  S/E  DESCRIPTION        DUR  NOV74     NOV74     DEC74     DEC74     JAN75     JAN75

                                  X....*....I....*....I....*....I....*....I....*....I....*....

A    1    SKETCH PLANS &      28  CCCCCCCCCCCCCCCCCCCCCCCCCCCCC    I    *    I    *    I    *
          ESTIMATES                x    *    I    *    I    *    I    *    I    *    I    *

1    2    OUTLINE TOWN        21  X    *    I    *    I    *    AAAAAAAAAAAAAAAAAAAAAA.......
          PLANNING                 X    *    I    *    I    *    I    *    I    *    I    *

1    3    CONSIDERATION OF    28  X    *    I    *    I    *    CCCCCCCCCCCCUCUCCCCCCCCCCCUCCCCC
          PLANS                    X    *    I    *    I    *    I    *    I    *    I    *

3    4    COMPLETION STAT.    28  X    *    I    *    I    *    I    *    I    *    I    *AAAA+
          APPROVAL                 X    *    I    *    I    *    I    *    I    *    I    *

3    5    SUBCONTRACT         42  X    *    I    *    T    *    I    *    I    *    I    *AAAA+
          DETAILS/ESTIMATES        X    *    I    *    I    *    I    *    I    *    I    *

3    6    CONSULTANTS         56  X    *    I    *    I    *    I    *    I    *    I    *AAAA+
          DRAWINGS                 X    *    I    *    I    *    I    *    I    *    I    *

3    7    WORKING DRAWINGS    98  X    *    I    *    I    *    I    *    I    *    I    *CCCC+
          *                        X    *    I    *    I    *    I    *    I    *    I    *

7    8    PREPARE BILLS OF    35  X    *    I    *    I    *    I    *    I    *    I    *
          QUANTITIES               X    *    I    *    I    *    I    *    I    *    I    *

8    9    PRINT BILLS OF       7  X    *    I    *    I    *    I    *    I    *    I    *
          QUANTITIES               X    *    I    *    I    *    I    *    I    *    I    *

9    10   TENDERING           28  X    *    I    *    I    *    I    *    I    *    I    *
                                   X    *    I    *    I    *    I    *    I    *    I    *

9    11   ANNOTATE BILLS OF    7  X    *    I    *    I    *    I    *    I    *    I    *
          QUANTITIES               X    *    I    *    I    *    I    *    I    *    I    *

11   12   EXAMINE PRICED       7  X    *    I    *    I    *    I    *    I    *    I    *
          BILLS                    X    *    I    *    I    *    I    *    I    *    I    *

10   13   CONSIDER TENDERS    21  X    *    I    *    I    *    I    *    I    *    I    *
                                   X    *    I    *    I    *    I    *    I    *    I    *

13   14   PREPARE CONTRACT     7  X    *    I    *    I    *    I    *    I    *    I    *
          DOCUMENTS                X    *    I    *    I    *    I    *    I    *    I    *

14   15   SIGN CONTRACT        1  X    *    I    *    I    *    I    *    I    *    I    *
                                   X    *    I    *    I    *    I    *    I    *    I    *
```

```
                    CCCC : CRITICAL  ACTIVITY
                    AAAA : NON CRITICAL  ACTIVITY
                    .... : FLOAT   TIME
```

manpower, materials, machines and capital. In order to undertake these tasks the computer must be fed with the following information.

(a) The resources required for each activity; these may be expressed as a rate, for example seven operatives required for each day of the activity. Alternatively the resources may be expressed as a total, for example 25 000 bricks required to accomplish the activity.

(b) The total resources available for each project; these may be expressed as a constant throughout the project or they may vary. For instance the

BAR CHART

6	20	6	20	3	17	1	15	29
FEB75	FEB75	MAR75	MAR75	APR75	APR75	MAY75	MAY75	MAY75

```
I....*....I....*....I....*....I....*....I....*....I....*....I....*....I....*....I....*....

I   *   I   *   I   *   I   *   I   *   I   *   I   *   I   *   I   *
I   *   I   *   I   *   I   *   I   *   I   *   I   *   I   *   I   *

I   *   I   *   I   *   I   *   I   *   I   *   I   *   I   *   I   *
I   *   I   *   I   *   I   *   I   *   I   *   I   *   I   *   I   *

I   *   I   *   I   *   I   *   I   *   I   *   I   *   I   *   I   *
I   *   I   *   I   *   I   *   I   *   I   *   I   *   I   *   I   *

-   AAAAAAAAAAAAAAAAAAAAAAAAA............................................................  +
I   *   I   *   I   *   I   *   I   *   I   *   I   *   I   *   I   *

-   AAAAAAAAAAAAAAAAAAAAAAAAAAAAAAAAAAAAAAAA..........................................  +
I   *   I   *   I   *   I   *   I   *   I   *   I   *   I   *   I   *

-   AAAAAAAAAAAAAAAAAAAAAAAAAAAAAAAAAAAAAAAAAAAAAAAAAAAAAAAAA...........................  +
I   *   I   *   I   *   I   *   I   *   I   *   I   *   I   *   I   *

-   CCCCCCCCCCCCCCCCCCCCCCCCCCCCCCCCCCCCCCCCCCCCCCCCCCCCCCCCCCCCCCCCCCCCCCCCCCCCCCCCCCCCCCCC.  +
I   *   I   *   I   *   I   *   I   *   I   *   I   *   I   *   I   *

I   *   I   *   I   *   I   *   I   *   I   *   I   *   I   *   I   *
I   *   I   *   I   *   I   *   I   *   I   *   I   *   I   *   I   *

I   *   I   *   I   *   I   *   I   *   I   *   I   *   I   *   I   *
I   *   I   *   I   *   I   *   I   *   I   *   I   *   I   *   I   *

I   *   I   *   I   *   I   *   I   *   I   *   I   *   I   *   I   *
I   *   I   *   I   *   I   *   I   *   I   *   I   *   I   *   I   *

I   *   I   *   I   *   I   *   I   *   I   *   I   *   I   *   I   *
I   *   I   *   I   *   I   *   I   *   I   *   I   *   I   *   I   *

I   *   I   *   I   *   I   *   I   *   I   *   I   *   I   *   I   *
I   *   I   *   I   *   I   *   I   *   I   *   I   *   I   *   I   *

I   *   I   *   I   *   I   *   I   *   I   *   I   *   I   *   I   *
I   *   I   *   I   *   I   *   I   *   I   *   I   *   I   *   I   *

I   *   I   *   I   *   I   *   I   *   I   *   I   *   I   *   I   *
I   *   I   *   I   *   I   *   I   *   I   *   I   *   I   *   I   *

I   *   I   *   I   *   I   *   I   *   I   *   I   *   I   *   I   *
I   *   I   *   I   *   I   *   I   *   I   *   I   *   I   *   I   *
```

resources may be expressed as

    4 assistants available between weeks 1 and 5
    2 assistants available between weeks 6 and 10

These may relate to the resources of the design team which comprise the staff of a particular profession.

(c) Any constraints imposed on the project by the management; these may relate to a time change in the execution of the project, which could vary the completion date given by the time analysis. If more time is allowed for the project the smaller will be the demand on resources. The facility permits the user to restrict the way in which a particular activity may be scheduled. This may involve the way in which the resources are split up or timed to start or finish consecutively.

A number of schemes may be performed using common resources in which event planning takes on a different outlook. Much time is spent compiling good operational schedules when dealing with complex projects and an even greater problem arises when multiproject planning and scheduling is required. Computerised facilities are available that enable schemes to be planned and scheduled on a multiproject basis whereby several projects can be related to each other. Any decisions taken with regard to a particular project can be measured to show their effects on other projects. The facility also allows subsequent simulations of alternative schedules. When a number of projects are involved the histograms of these may be superimposed and over-all requirements obtained over a given period. In the case of the quantity surveyor's own requirements the facility assists the organisation of the office manpower to suit the timing of projects. Staff requirements may also be checked against a target of staff time calculated from the fees for the work and that can be economically applied to each project. With the exception of an imaginative treatment when manipulating the data and the variable facilities contained in the programs, the process of multiproject scheduling is a comparatively simple process. While the computer will carry out the instructions as contained in the programs, it is the human element that ensures that decisions are based on sound reasoning. This means that the solving of this type of problem is an art and must be undertaken with professional skill and judgement.

Cost Control

It is important for large contracting organisations to adopt a suitable system of cost control to determine the planned and actual expenditures of a project. Network costing techniques form one of the best approaches to the cost planning and tracking of a project. Computerised facilities are available to provide management with a tool to analyse project costs and to keep a strict control over project expenditure. The facilities provided by International Computers Ltd in their PERT 1900 Series computer programs aim at providing a type of control that is directly linked with a network.[4] This is also considered to be a method of direct costing. Unfortunately the advance network techniques offered by the computer are not applied to the extent at which they would be most beneficial.

The application of costing to network programs forms an extension of the schedule control methods, which permit the evaluation of the current and future financial standing of projects. This provides an effective method of monitoring the costs and detecting any excess expenditure. Cost solutions can therefore be examined at an early stage in order to alleviate any uneconomic factors concerned with the project design and construction.

Cost analysis print-outs are similar to those obtained from the resource analyses, which may be in a numeric or combined numeric and graphic form. An example of the type of print-out available is shown in table 10.7. This shows the cost output in a combined numeric and graphic form. The headings are listed as follows.

(a) *The calendar date* – this indicates the day to which the line refers (DATE).
(b) *The planned cost* – this represents the cost of the project for the time period indicated and shows a cumulative total (C).

Table 10.7   Histogram print-out – cost analysis

```
COST OUTPUT                              RUN1    TIME NOW 14NOV74    PAGE 1

                                  0     10000    20000    30000    40000    50000

DATE       C      A    DIFF   V   I       I        I        I        I        I TIME
7JUL75    980    980    0         IA      I        I        I        I        I   .1
8JUL75   1960   1960    0  1010   IVA     I        I        I        I        I   .2
9JUL75   2940   2940    0  2020  I vA     I        I        I        I        I   .3
10JUL75  3920   4130   210 3240  I  VA    I        I        I        I        I   .4
11JUL75  4900   5130   230 4120  I   VA   I        I        I        I        I   .5
14JUL75  6110   6360   250 5270  I    VA  I        I        I        I        I  1.0
15JUL75  7320   7710   390 6580  I     VPA I        I        I        I        I  1.1
16JUL75  8530   9000   470 7890  I      VAI        I        I        I        I  1.2
17JUL75  9740  10330   590 9010  I       vA        I        I        I        I  1.3
18JUL75 10950  11630   680 10030 I        VPA      I        I        I        I  1.4
21JUL75 12290  13000   710 11370 I        IVPA     I        I        I        I  2.0
22JUL75 13630  14350   720 12760 I        I VvA    I        I        I        I  2.1
23JUL75 14970  15820   850 14240 I        I  VPA   I        I        I        I  2.2
24JUL75 16310  17230   920 15690 I        I   VPA I        I        I        I  2.3
25JUL75 17650  18580   930 16970 I        I    VPAI        I        I        I  2.4
28JUL75 18830  19700   870 18130 I        I    VvPA       I        I        I  3.0
29JUL75 20010  20790   780 xxxxxxxxxxxxxxxxxxxxxxxPAxxxxxxxxxxxxxxxxxxxxxxxxxxxxxxxxxx 3.1
30JUL75 21190  21950   760 I        I        IPA      I        1        I  3.2
31JUL75 22370  23150   780 I        I        I PA     I        I        I  3.3
1AUG75  23550  24360   810 I        I        I  PA    I        I        I  3.4
4AUG75  24760  25590   830 I        I        I   PA  I        I        I  4.0
5AUG75  25970  26910   940 I        I        I    PA I        I        I  4.1
6AUG75  27180  28150   970 I        I        I     PA I       I        I  4.2
7AUG75  28390  29260   870 I        I        I      PAI       I        I  4.3
8AUG75  29600  30440   840 I        I        I       PA       I        I  4.4
```

```
        C : CUMULATIVE  PLANNED  COST
        A : ACTUAL  COST
        V : VALUE  OF  WORK  DONE
     xxxx : TIME  OF  VALUATION
        P : PLANNED  COST
```

129

Table 10.8   Financial status report

REPORT DATE	PROJECT NAME	REPORT MONTH
29JUL75	CONTRACT 046	1
	STATUS REPORT	

EXPENDITURE				VALUE		OUTLOOK					
AUTH	ACT	DIFF	TOTAL	TOTAL DIFF	AUTH SUM	COMMIT	PER CENT	TOTAL AUTH SUM	DIFF SINCE AUTH	PER CENT	OUTLOOK COST
A	B	C	D	E	F	G	H	J	K	L	M
18830	19700	+870	18130	-1570	105300	111870	106	125000	+5000	101	131570

(c) *The actual cost* – this is a report of the actual cost at the time to which the line refers (A).

(d) *A histogram of the cost involved* – this graphic representation of the costs forms a cost curve that indicates (i) the planned cost (P), (ii) the actual cost (A) and (iii) the current value of the work executed (V).

(e) *The number of weeks from the base date* – this represents the number of weeks that the project has been performed and is shown in weeks and days. For example, 4.2 represents four weeks and two days.

A basic requirement of cost control is the prediction of future expenditure. This is usually satisfied by the production of financial reports and predictions. The predicted costs are shown below the 'time now' line, which is indicated by a line of 'x' marks (table 10.7). This information is sometimes produced in a report of the financial position of the project, which is also a print-out by the cost analysis programs. These present the financial position of a project at the time of reporting.

An example of a financial-status report is shown in table 10.8 and it usually contains information under the following headings.

(a) The original total estimated cost of the project (J:£125 000)
(b) Details of any planned additional costs of variations and the like (K:£5000)
(c) The original planned estimated cost of the project at the time of reporting (A:£18 830)
(d) The value of work performed by activities in the network (D:£18 130)
(e) The actual cost of the project at the time of reporting (B:£19 700)
(f) Details of the planned cost of completing the project. This is the difference between the overall planned costs and the actual value of work performed by activities (G:£111 870)
(g) A prediction of the total project cost at completion (M:£131 570)

Updating

The network analysis prepared by a computer is considered to be justified only if it is linked to a regular updating routine. Progress should be maintained as closely as possible to the planned program of work. In order to achieve this it is necessary to make comparisons between achievement and forecast during the progress of the work. As situations change and events become nearer it may be necessary to reassess and adjust the program of work. Continual adjustment and updating of the program is therefore necessary and in order to maintain an up-to-date appraisal of the continuously changing situation the computer must be supplied with data as follows.

(a) The reporting date
(b) The activites completed
(c) The activities in progress with an estimate of the time required for completion
(d) Revision of estimates of the duration of the activities not started
(e) Any necessary revisions of network logic

This information is input to the computer, which incorporates it with the main network and produces a revised analysis. The updated network analysis will identify the areas where action is required by management.

FINANCIAL FORECASTS, INTERIM CERTIFICATES AND FINAL ACCOUNTS

The quantity surveyor's post-contract services involve periodic valuations for the purpose of financial forecasts and interim certificates. The computer may be programmed to produce progress cost control reports using the contractor's prices and the job file. This facility permits the introduction of any type of adjustment, which may take the form of measured omissions and additions, possibly involving complete remeasurement of whole sections of work. From this information the value of the project may be obtained at any time during building operations.

It is possible for the computer to produce interim certificates from input data consisting of the valuation of individual items or unit quantities. Payments on account are generally based on the valuation of sets of bill items that are grouped into building activites. This may mean the regrouping of items into activity bill formats. Certain techniques involve the preparation of a valuation grid, which consists of a number of computer storage locations each carrying a classified value of an element of work held by the computer.[6] A valuation may thus be produced by providing details of the percentage value of work carried out on site relating to each location. Each interim certificate is based on a statement of account, which is the computer print-out. This may be made available in the form of a summarised statement or a detailed priced statement, and the latter may be obtained with or without descriptions. Interim valuation reports may also be produced in any of the following: (a) an expenditure flow forecast based on the latest information available, (b) a comparison of estimated with actual expenditure and (c) a schedule of items with descriptions to which rates may be fixed.

Final accounts are based on the valuation of work based on a bill of variations. This is prepared on an add/omit basis unless considerable remeasurement is necessary. In the case of remeasurement contracts a detailed print-out can be produced as the measurements and rates for each item are agreed. These are collectively built up by the computer and a final account is produced in full in a bill of quantities format. In the event of an abnormal termination of the contract, such as the insolvency of either party, the computer can produce details of the work not performed. The quantities and descriptions may then be presented as a new tender document. This is a useful facility and the document may be used for negotiating the completion of the work by a new contractor.

COORDINATED INFORMATION SYSTEMS

The information produced by the quantity surveyor throughout the building process will have a greater significance if it is used to support management techniques. This involves a system of data coordination as described in chapter 6. For this purpose computers can be used effectively for solving problems associated with the need for control and economy at every stage of the building process. Computer programs have been developed to provide coordinated information systems for use during the design and construction processes. For instance a system produced by Construction Control Systems Ltd (CCS) may be used for any of the following purposes.[7]

(a) Schedules of quantities in any item sortation
(b) Financial forecast and control
(c) Resource analysis of design and production
(d) Materials ordering
(e) Labour and plant resource allocation
(f) Network analysis
(g) Construction method analysis
(h) Bonusing
(i) Specifications

The applications of this system embrace the organisation and presentation of the design brief, specifications, drawings, technical information and correspondence, and they have been used for the management of many kinds of project such as housing, shopping centres, office blocks, hospitals and schools. The system can be applied as a series of master files for use during one or all the stages of the building process. Alternatively it can be used as a separate application at any stage of the building process using individual files relevant only to that stage. A series of manuals and instructions enables the user to become familiar with those parts of the system that particularly affect him. All that is required is a knowledge of the simple rules relating to the completion of the input forms and the code structure as described in chapter 6.

REFERENCES

1. —— Computer System Handbook, Part 3M, Consortium of Local Authorities Special Programme: Central Development Group.
2. —— Computer Aided Quantity Surveying: a Comprehensive System, Central Electricity Board Transmission Division.
3. —— A Computer Based Cost Analysis System, Construction Control Systems Ltd.

4. —— *ICL PERT Users Guide 1900 Series* (Technical Publications Service, ICL, London, 1969).
5. W. R. Martin, *Network Planning for Building Construction* (Heinemann, London, 1969).
6. K. W. Monk and T. D. Dunstone, 'Two Years of Practical Quantity Surveying by Computer', *Chart. Sur.*, 98 (1966) pp. 363–8.
7. —— General Brochure, Construction Control Systems Ltd.

11 COMPUTER SERVICES

Computerised data-processing has brought the quantity surveyor into contact with specialists whom he would not previously have met in the performance of his duties. In this situation the quanitity surveyor needs to know certain facts about the computer industry that will enable him to make a choice regarding the best way of utilising the computer. It is important that he have some knowledge of the computer industry and the services it provides. This chapter is therefore concerned with an examination of the computer industry and the methods of utilising the various services provided by the industry.

While the computer industry is dominated by a number of companies producing main-frame computers, there is no shortage of other firms in business each providing various types of computer service. Apart from the manufacturers of hardware the industry operates to provide (a) *consultancy services* and (b) *processing services*. Main-frame computer manufacturers have always had differing attitudes towards the services they provide. Certain manufacturers provide a 'package deal' whereby the customer buys a package consisting of a computer and peripherals with software applications and programs to serve the user's particular needs, including staff education to put the system into operation. Other manufacturers deliberately confine their activities, which then have to be supplemented by service companies who offer some specialised services in the running of a computer system. This means that the provision of systems designs and software must be provided by agencies offering a consultancy service.

CONSULTANCY SERVICES

The quantity surveyor, when considering using a computer system will require expert advice on the various possibilities open to him. This may be obtained from such sources as computer consultants, or programming agencies and software houses.[1]

Computer Consultants

This type of service concerns those firms who do not sell computer time or computer software, but rather brain-power. They will advise on all the services relating to a computer installation and management. Their services also include such things as computer organisation and planning, giving advice on computer investment, design of data communication systems and the like. Any equipment they possess is solely for their own use in the performance of their services.

The decision whether to use a consultant or not can only be made by considering the particular circumstances. There are some processes that are well suited to a consultancy exercise, such as the technical aspects of choosing hardware, selecting applications, staff training and feasibility studies. Generally speaking the reason for using a computer consultant lies in the fact that he is able to offer specific technical advice with a knowledge and depth of experience that are not available to the quantity surveyor. Some of the reasons for seeking the services of a computer consultant may be summarised as follows.

(a) Feasibility studies — the purpose is to provide unbiased advice on whether or not a computer should be used and to provide a plan for all that follows.
(b) Equipment evaluation — this evaluation will enable the right choice of machine, bureau or terminal best suited for the circumstances.
(c) Recruitment and training — useful advice may be obtained on the selection of staff and provision of suitable training for the staff.
(d) Systems design — this is of crucial importance and the consultant can act as coordinator between user and computer professional.
(e) Programming — this is not just a matter of sending a man on a programming course, since it takes six to twelve months to reach a satisfactory standard. It is during this period that the consultant can offer experienced guidance.
(f) Implementation — at this stage the consultant can draw up a realistic plan covering such things as systems design, programming, site preparation, recruitment and training.
(g) Computer audits — an assessment of the current efficiency of computer operations may be required with a possible evaluation of plans for future action.

The services performed by a computer consultant may be likened to the services of an architect.

Software Houses and Programming Agencies

In general any collection of programs that can turn the computer into an effective tool is termed *software*. The term is often used specifically to describe general-purpose programs such as assemblers and compiling routines, as opposed to those programs that are written for a computer user for his own purposes. Many management consultants and a few service companies develop software for their customers. The more sophisticated software for business and industry requires extensive and highly specialised talent that can only be provided by this type of agency. Software houses and programming service companies will write any program to suit the user's needs. Software houses range from the large multipurpose companies to the small supplier who specialises in a single industry or application. The services provided may be considered under the following headings.

Consultancy   This includes the type of services provided by computer consultants. In addition many software houses provide workshop training linked with the use of a particular computer system. Courses on computer applications and techniques may also be provided by these agencies. This is a particular area where an agency may be useful in the recruitment of technical staff.

Implementation   A large proportion of the work of the software house lies with the implementation of computer systems. The services provided under this heading include the systems analysis and design, or programming services to replace the internal programming resources of an organisation.

Management   The services provided under this heading include the following types of management.

(a) Caretaker management pending the establishment and organisation of a computer department.
(b) Services management involving the continuation of external management which leads to the phasing out of the user's responsibility.
(c) Facilities management which involves the management of computer facilities by external agencies who take over the computer or supply computer time.

An important consultancy service required by the quantity surveyor is the implementation of computer systems and in particular the provision of programs for processing his data. For this purpose various agencies have developed *application packages* for quantity surveyors' use.

APPLICATION PACKAGES

An application package is a suite of programs designed and written for a particular problem such as the preparation of bills of quantities. A number of application packages has been developed by various agencies for a restricted specified use or for general use by quantity surveyors. The origin of an application package may stem from a number of sources depending on the problem situation and the means available for developing a computer processing system. These may be considered to stem from any one of the following sources.

(a) A company may be established whose object is solely the development and marketing of an application package for general use. The package may be offered for sale to be used by a quantity surveyor with his own computer or to a computer bureau for use by them as a means of providing computer processing services. A typical example of this type of package is marketed under the name of Construction Control Systems Ltd (CCS).
(b) Computer systems may be developed for their own use by various consortia, such as Local Authorities, with mutual interests and problems who pool their knowledge and resources. The coordination of efforts in the development of computerisation by local authorities is carried out by an organisation known as LAMSAC.
(c) Software houses and service bureaux may develop packages according to the particular requirements of a user. These may be developed for personal use initially but subsequently may be marketed for general use when the system has been proved. A number of such packages has been developed in this way and close liaison has been maintained between surveyors and this section of the computer industry. Some of the better-known packages are AUTOBILL, COSYBILL, DGS and FACET.
(d) A number of general-purpose packages has been developed by subsidiary agencies of main-frame manufacturers for general use, an example being the ICL 1900 Series PERT programs.

The nature and characteristics of the foregoing packages are examined hereafter in a little more detail to illustrate their scope and methods of use.

Construction Control Systems Ltd (CCS)   A computerised information system is offered by Construction Control Systems Ltd for use in the various stages of design, documentation and construction processes.[2] The CCS system is based upon the CBC system of contract documentation. This is a management system organisation that provides a method of coordinating information for the building process on drawings, bills of quantities, specifications, catalogues, site reports, work studies, invoices and accounts (see chapter 6). Thus the CCS system is a computerised method of applying the CBC system of contract documentation. The activities of the CCS system are concerned with the application of computer techniques covering such purposes as

  (i)   bills of quantities in any sortation
 (ii)   financial forecasts and control
(iii)   resource analysis of design and construction
 (iv)   materials ordering
  (v)   labour and plant resource allocation
 (vi)   network analysis
(vii)   construction method analysis
(viii) bonusing
 (ix)   specifications.

Programs have been developed by CCS that can be purchased, leased or hired. Purchasing will be with the object of using the system on an in-house computer. In this instance CCS will undertake both the responsibility of setting up the package for operational use and the training of staff involved in its operation. CCS will also lease the package on the basis of a fixed annual charge over an agreed number of years. This particular arrangement maintains a continuing relationship with the company in circumstances where an organisation feels it to be appropriate. The system may be used when an organisation does not possess any computer facilities and in such a situation the package may be made available on hire and CCS will establish the system in collaboration with a computer service bureau. The system can be applied either as a series of master files for use during one or all the stages of the building process, or as a separate application for use at any stage of the building process with files relevant only at that stage.

CCS Ltd was established in Great Britain under the name of Coordinated Building Communications Ltd (CBC) in 1964, offering a consultancy service for the implementation of the CBC system as developed in Denmark by Bjorn and Knud Bindslev. The company was restructured in 1971 to integrate all the skills associated with construction projects together with the capacity for electronic data-processing. The name of the company was changed to its present form in 1971 in order to reflect the scope of the company's activities more accurately. The company provides a consultancy service offering advice on electronic data-processing at all stages from initial feasibility study to implementation of the system. A development section is also available to write a detailed system specification together with additional programs that may be required.

LAMSAC   Local Authorities Management Services and Computer Committee is an organisation whose constituent body consists of various committees, which are (i) a computer panel of LAMSAC, (ii) a building construction applications group, (iii) a bills of quantities working party. Each committee comprises representatives of local authorities. The LAMSAC computer system is offered to quantity surveyors for in-house use on any one of a number of types of computer. The implementation and development of the system rests with a number of 'sponsoring authorities' who are responsible for ensuring that the users of the system are kept informed of program amendments and updates relating to a particular type of computer. Each sponsoring authority is responsible for the operation of the system on the particular type of computer with which it is familiar.

Many sponsoring authorities hold users group meetings that provide an opportunity for member users to discuss problems and put forward development suggestions. These would then be passed on to the LAMSAC Steering and Development Group of the computer panel with the object of maintaining and updating programs and developing future work. This obviates the duplication of effort and maintains an up-to-date library and suite of programs. Liaison between the computer personnel and the sponsoring authorities is maintained through a Computer Managers Working Party.

The basis of the LAMSAC computer system is the LAMSAC library of descriptions. The embryo of the whole system emerged in May 1965 as a suite of programs that had been developed by Hertfordshire County Council to produce bills of quantities by computer. It was in April 1966 that the Consortium of Local Authorities Special Programmes (CLASP Development Group) adopted this system with a view to development for its own use. One of the first tasks was the rewriting of the Hertfordshire standard library to conform to Fletcher/ Moore standard phraseology. This was prompted by a policy statement issued by the Royal Institution of Chartered Surveyors, which recommended that attempts should be made to accommodate the Fletcher/Moore standard phraseology in the application of computers to quantity surveying. The library was subsequently revised by a joint CLASP/SEAC working group in December 1967. LAMSAC commissioned the preparation of a metric version of this library, after financing a technical edit by Leonard Fletcher and Partners, for its suitability and use on a national basis by Local Authorities. This was completed in May 1969 and the metric edition of the LAMSAC library of descriptions was subsequently published in 1970.

The computer programs that support the LAMSAC library offer the following facilities.[3]

(i) A basic suite for the preparation of bills of quantities with additional features for short coding and unit quantities and updating facilities.
(ii) Cost analysis.
(iii) Consolidated cost index for use in cost planning when pricing elements.
(iv) Automatic pricing of bills of quantities.
(v) Short code repricing for updating by a single adjustment possibly on a percentage basis.

The system is recommended by LAMSAC as a step towards an integrated management system for the building industry, and development work is being carried out in the field of standardised cost planning techniques, computer-aided building design programs, standard construction programs, network analysis, resource schedules and other necessary aids to construction.

AUTOBILL   This is a computer system for the production of bills of quantities, cost analyses, final accounts and other associated reports. Cost analyses may be produced in a varying number of sequences when rates have been established.[4] Also pre-tender priced elemental abstracts and post-tender priced abstracts may be produced using the quantity surveyor's estimated rates or the contractor's actual rates respectively. A locational cost analysis may be provided where different parts of working drawings have been coded. In the case of negotiated contracts a budgetary analysis of costs may be produced relating to the contractor's own major building operations. The system is operated by Centre-file (Northern) Ltd, Manchester, a member of the National Westminster Bank Group. Rather than offer a full range of services covering a variety of applications they have elected to limit their services to providing specific applications, one of which is AUTOBILL.

The system is operated mainly by batch processing with punched paper tape input but facilities for remote batch entry may be provided if so desired. The punched paper tape is prepared by the quantity surveyor on an automatic typewriter. This is a machine with added facilities for producing punched paper tape. The tapes are sent to the bureau in suitably sized batches for processing. The system provides approximately one hundred consistency tests to detect errors. These are output from the computer in the form of an errors report, which lists all the data received with brief commentaries and coded explanations against the rejected data. All errors must be corrected, repunched and forwarded to the bureau with subsequent batches of punched data. When all the correct data have been received the bureau will print an abstract.

Standard libraries of descriptions, based on Fletcher/Moore standard phraseology, are held on magnetic tape in the computer centre. However, these may be held on edge punched cards by the quantity surveyor and although standard libraries are available for use with the system the quantity surveyor may create his own. These may be written in any language based on Fletcher Moore or any other phraseology. A library may be created by coding a previously completed bill, which then becomes the code manual. By merely using page and item number references this may be adapted to the short coding techniques. The system permits the production of a descriptive bill using rogue descriptions for all the items in the bill. The system may also be used without any item description by processing and sorting items using codes inserted by the taker-off in lieu of descriptions. The print-out is a list of codes in the form of an abstract.

A draft bill is produced on the computer's printer for editing purposes, and on this last minute alterations may be made. A punched paper tape version is automatically produced in addition to the printed document. The quantity surveyor may produce either stencils or offset masters for final bill duplication by passing the punched paper tape version through the automatic typewriter. The machine may be stopped and last minute amendments inserted manually. This method of processing provides the quantity surveyor with complete control up to the last minute of production.

COSYBILL   County Surveyors' Society SYstem for BILLs of quantities is a system for producing engineering bills of quantities in accordance with D.O.E.

specifications.[5] The system originated in the County Surveyor's Department of Buckinghamshire County Council and was developed by a working party set up by the ICL Users Group of the County Surveyors' Society. Programs have been written covering all stages of bill production from taking off to the comparison of tenders and production of interim certificates. COSYBILL contains standard libraries for use with engineering projects such as road and bridge works and it is in this area that the package is most beneficial. However, it may be used for construction work generally with a revised library.

DGS   The DG System was developed by a computer bureau, Centre-file (Northern) Ltd, in accordance with the requirements specified by a Development Group of Chartered Surveyors.[6] It was launched on 1 January 1973 as a system for the production of bills of quantities. The system is an attempt to assist and accelerate the taking-off process and reduce the manual processes of measurement and working up by transferring certain of these manual procedures to computer processing.

The system is designed to reduce the amount of written entry by taking off directly in code on specially designed take-off sheets using a technique of automatic measurement. This involves direct coding by the taker-off at the same time that he enters the dimensions, where the introduction of a traditional coder would tend to restrict the automatic capabilities of the system. The automatic measurement process forms an integral part of the system as a prerequisite to the operation of the system. It is therefore important that the taker-off understand the automatic facilities contained in the system and adapt his measuring techniques to take full advantage of the processing facilities. The automatic facilities contained in the system provide for the following.

(i)   The automatic classification of items based on dimensions.
(ii) The automatic generation of items that relate to the presence of measured items.
(iii) A system of repeat coding.

The *automatic classification of items* is based on the dimensions entered on the input sheets and the computer will automatically generate the various classifications of descriptions and units of measurement. For instance, the following classifications are required by the Standard Method of Measurement depending on the dimensions of the item.

(1) Excavations and planking and strutting are classified according to their depths in 1.50 m stages.
(2) Surface excavation, hardcore and concrete beds have their units of measurement classified according to their thicknesses over 300 mm deep in cubic metres and not exceeding 300 mm deep in square metres.
(3) Concrete in foundations is classified according to its thickness in stages not exceeding 150 mm thick, over 150 mm but not exceeding 300 mm thick, and over 300 mm thick.

The system of measurement requires a single code entry for the item with a varying number of lines for measurements. The code entry for descriptions is based on codes contained in a coding manual. The manual is simply an expanded standard bill consisting of coded items using a code structure that utilises the page number and the item reference on the page. This is a familiar look-up pattern that most quantity surveyors use. A code for an item may appear thus: 15/9 A1 for say 'Concrete 1:2:4/40 mm aggregate in foundations'.

Measurements in accordance with the traditional methods may include the following.

42.65 0.75 0.12	Concrete 1:2:4/40 mm aggregate in foundations not exceeding 150 mm thick
20.65 0.89 0.15	
60.75 0.90 0.23	Ditto over 150 mm but not exceeding 300 mm thick
2/ 23.45 0.90 0.32	Ditto over 300 mm thick

The coded entries would be inserted thus

15/9 A1	42.65	0.75	0.12
	20.65	0.89	0.15
	60.75	0.90	0.23
	2/ 23.45	0.90	0.32

Dimensions are input in their raw state and are squared by the computer which refers to the third dimension and groups the quantity in the appropriate description classification.

The *automatic generation of items* means that the computer will pick up any associated items relating to measured work. For instance, certain protection clauses will be generated by virtue of the presence of measured items relating to the sections in the Standard Method of Measurement requiring such a clause. Clauses may also be picked up in relation to measured excavations, such as keeping all excavations free from storm or percolating water and associated items of spoil disposal. The latter would be adjusted with each subsequent measured item for the reuse of spoil such as filling. The foregoing items would generally need to be picked up either as part of the take-off or at the working-up stage. The system therefore relieves the taker-off or worker-up of the additional task of having to account for the items.

Repeat coding techniques are also included in the system of processing, whereby item codes may be repeated with new dimensions. Also complete pages of item codes and dimensions may be repeated with a new timesing factor. Conversion factors may be introduced that will cause the computer to convert dimensions into weights that may be required for such items as steel reinforcement or steel sections. A facility is also provided for the interchange of items between trade sections. This means that items may be extracted from their original trade classification to appear in print-outs in another trade classification. This facility enables *ad hoc* combinations of items to be produced without any duplication of descriptions or codes within the library.

The system contains a facetted library, which eliminates phrases and codes for whole items or parts of items that may be generated by the computer from other input. The library is based on the Standard Method of Measurement of Building Works and may be adapted to suit any subsequent revision of the document. The advantages of the system claimed by its users are as follows.

(a) The compactness of the library in terms of its ratio of coverage of items.
(b) Easy referencing and the comprehensibility of the coding.
(c) The simplification of the taking-off process in terms of the processing techniques, which provides for automatic measuring and the automatic coverage of specification details.
(d) The processing facility that provides an interchange of items between work sections.

Input consists of elemental and locational references, coded description references and unsquared dimensions on punched paper tape, which is sent to the bureau for processing. The system allows batches of data for various projects to be fed into the computer simultaneously. After processing they are stored in appropriate batches providing cumulative totals for each project. When processing is complete the data are transferred to a completed job file where they are kept for further post-bill processing. An error listing and an optional data listing are produced as a means of checking input data and approving all corrections. When all taking-off is complete and all errors have been corrected the computer will produce all the required output documents automatically. This is generated by an appropriate command signifying that the final batch of data has been received for that project and contains the necessary set of parameter codes indicating the form of output required.

Final output consists of abstracts, cost analyses and bills in any sortation that may be priced. An abstract or draft bill is produced on line and may be used for checking and editing before the final bill is produced. An 'in clear' listing may be produced which lists all the free descriptions including rogues. This focuses the editing on the non-standard phraseology and provides a means of correcting and unifying the varying descriptions produced by takers-off for identical items. Final bill production varies according to the choice of method, which may involve (i) photographic reproduction of the computer's line printer copy suitably reduced in size, (ii) producing a printed bill copy on an automatic typewriter using a paper tape version of the bill that has been produced as a by-product of processing, (iii) retyping a master stencil from the abstract or the draft bill for lithographic reproduction.

FACET   Faber Cost Estimating Technique is a computerised system for the preparation of cost estimates, cost analyses and bills of quantities. It was developed by Oscar Faber and Partners for their own use in connection with structural and environmental engineering.[7] However, it is intended that the system should be equally used for building quantities. The system is designed to produce bills of quantities and to provide comprehensive and detailed cost planning studies, which may be carried out coincidentally with design.

The system is operated through the IBM terminal business system, which involves the use of a keyboard terminal through a Datel service to a computer that provides on-line facilities. A variable number of terminals may be used that permit access to common data. It is also possible for the interchange of information between terminals. The method of operating the system is by *remote job entry* to the computer centre, which processes the data in batches. Access to a file is immediate through nominated terminals. Subsequent processing may be completed within two hours in the normal batch queue or two to fifteen minutes if the fast processing facilities are used. The system provides an error check on the data transmitted with an error report while the operator is still on line. This is a conversational time-sharing facility and any errors discovered may be corrected immediately. An audit report is subsequently produced that contains all the accepted items. These are also recorded on a work item file. The audit report provides a visual check on the data keyed in, which makes it possible at this stage to revise any part of the take-off or re-enter any incorrectly coded items that have been rejected.

The system uses three master library files that rely on prepared information relating to standard phrases, descriptions and rates. The editing process involves the computer gathering relevant phrases or descriptions from the master files and adding the rates from the master rates library before extending the items for audit. It is intended that the system should have a general application to the building process as well as its apparent engineering applications. The coding has therefore been designed to suit building components for a building work section bill.

The major output documents consist of a take-off audit report and bill reports, which provide priced and unpriced bills and collection sheets. The bill reports are printed on the computer's line printer and are subsequently duplicated on a Rank Xerox machine that reduces the documents to A4 size sheets. Bills may be printed in upper- or lower-case characters at an extra cost. This may be considered justified by the improved appearance of the bills.

PERT   International Computers Ltd have produced a package known as 1900 Series PERT Package for magnetic tape orientated configurations.[8] This is made available by a number of computer service bureaux providing specific program facilities that may be used during the processing of a complex mix of networks. The program contains features that produce time analyses, library networks, selective progress control, resource analyses, multiproject scheduling and cost control (chapter 10).

PROCESSING SERVICES

An alternative to acquiring a computer and directly employing data-processing staff is for the user to enlist the services of a *computer service bureau*.[9] This is an agency that provides services in the specialised field of electronic data-processing and may be obtained from a number of different sources as follows.

(a) Computer manufacturers with a separate organisation for providing computing facilities for data processing.
(b) Computer users with their own computer who have spare capacity.
(c) Independent companies specially formed for the provision of computing services to clients.

A service bureau provides various methods of operation whereby the client may use the bureau's own computer to run a job on the machine himself. The operations may be carried out by the client with or without the bureau's operators in attendance. This is a 'do-it-yourself' service that is usually available during off-peak periods. Another system involves hiring computer time for processing the clients' data by the clients' own programs. Hire charges vary according to the time of day and the length of time the facilities are provided; evening processing, for instance, is cheaper. Service bureaux may also offer a full range of software including processing by application packages. These services involve the preparation of raw data by the client, which are processed as a complete package by the bureau.

One of the main problems in using the systems provided by a bureau is the degree of quantity surveying involvement. Between the taking-off procedure and the data processing there are a number of areas of responsibility that can rest with either the quantity surveyor or the bureau. These are considered to be (i) coding, (ii) punching and verifying coded dimensions, (iii) production of a master copy of the final bill pages, and (iv) printing the bills of quantities.

A number of trained and experienced coders may be employed by a service bureau and made available for punching their customers' work. However, they normally supplement the efforts of the quantity surveyor when he is particularly busy. The quantity surveyor is advised to perform the coding operation himself since the bill descriptions are determined at this stage and the responsibility for this should remain with him.

Punching is another operation that the quantity surveyor may elect to perform. This would mean the acquisition of pieces of equipment, possibly in sufficient numbers to avoid problems in the event of a breakdown. The use of such equipment is seen to be justified when there is a high volume of work and using it in this way means that the dimensions remain with the quantity surveyor and never leave his office. Usually a reduction in the processing fees will be made if this operation is performed by the quantity surveyor. In some instances the preparation of punched data may involve an *automatic coding technique*, which utilises a standard library on edge punched cards. The taker-off describes the items so that they correspond with the standard descriptions in the library. Each item is then punched by the operator selecting a matching edge punched card containing the description of the item from the library file. The card is put into a machine that automatically types the description and punches the correct code for input. The edge punched card causes the machine to tabulate the description to a preset position and stop. At this point the dimensions are punched. The machine will print the figures for verification while being punched. Having satisfactorily completed this procedure the operator selects the next edge punched card for the next description.

A master draft copy of the bills of quantities must be produced before the bills are printed. This sets out the contents of the bill to establish the final bill page formats. A method of achieving this is by the use of a paper-tape-driven typewriter. The paper tape may be a by-product of a computer run and may be used with an automatic typewriter as an off-line method of printing. The cost of owning this type of device may, however, prohibit the quantity surveyor performing this operation himself, and better results may be obtained with the use of more advanced techniques than the bureau can provide. If the quantity surveyor does not possess his own printing facilities the bureau may print a number of copies suitably bound.

The security of data is an important consideration. The bureau will usually require all grades of staff to sign a specific declaration requiring them to treat their customers' information in the strictest confidence. At the same time internal security provides that neither the client's name nor project names or sizes be disclosed; references are usually made by means of job numbers rather than names. Bureau services may involve batch processing, which means that input data have to be transferred from the quantity surveyor's premises to the bureau's premises. Sometimes a bureau will provide a courier service to facilitate an efficient and speedy transfer.

The decision to use a service bureau may stem from a number of different reasons.

(1) The bureau services may be used to supplement in-house computer facilities. A decision might be made to obtain the services of a computer bureau for a limited period in order to gain valuable experience in computing before deciding on whether to provide in-house facilities. The facilities provided by certain bureaux enable the testing and running of programs before being used on a similar in-house installation.
(2) Consideration may be given to the work content — a major reason for selecting a bureau being that there is insufficient work to warrant the acquisition of a computer. Bureaux may be used to provide certain standby facilities to back up an in-house computer in the event of a breakdown. At the same time they may be used at peak periods as a back-up, particularly where the in-house facilities have insufficient capacity.
(3) More beneficial facilities may be gained by using a bureau that could offer the use of equipment that is more up to date and more powerful using efficient techniques.
(4) It may be more economic.

Computer service bureaux have formed an association known as The Computer Services and Bureaux Association (COSBA) to provide a code of practice for its members. There are two grades of membership depending on whether (a) the member is an experienced and established company operating a computer service bureau or software house (full member) or (b) the member is a small company concerned with the associated fields of services such as data preparation (associate member).

REMOTE COMPUTING

A useful facility that enables the quantity surveyor to use the computer is the technique of remote computing.[10] This facilitates the transfer of data through a

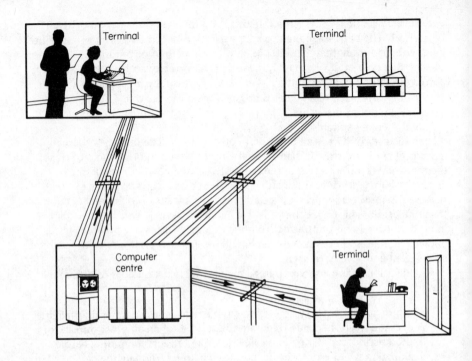

Figure 11.1   Remote computing

communication terminal for remote on-line processing. These techniques may be used in conjunction with in-house facilities or bureau services. Whichever service is provided will depend on the flexibility of the application package for use in conjunction with the computer facilities available.

A number of bureaux specialise in remote computing and provide facilities that involve the use of terminals situated in the user's premises or nearby (see figure 11.1). These facilities are designed to give a service to users who do not have their own computer installation and to others who find that certain work can be performed more economically in this way. The services provided by bureaux for remote on-line processing are known as *on-line bureau services.* Bureaux that do not provide on-line facilities are usually known as *batch bureau services.* Although they do not provide facilities for remote on-line processing they may possess facilities for remote off-line data transmission equipment.

The method of processing varies according to the facilities available at the bureau and on-line bureau services may be considered to fall into any one of the following categories.

(a) Remote batch entry (RBE)
(b) Remote job entry (RJE)
(c) Time sharing (TS) conversational interactive computing

Remote Batch Entry   Both the programs and the data are transmitted via a terminal to the central processing unit at the bureau. The input is generally punch card, paper tape or magnetic media but can be a keyboard entry or visual display. The information so transmitted is stored at the computer centre and batch processed when the computer is available, likewise the output from the processing is recorded for subsequent transmission back to the user.

Remote Job Entry   This method is similar to the remote batch entry method of processing but is on a smaller time-scale. The service is performed in real time. The user communicates on line direct to the central processor and enters his programs and data after invoking a language compiler. The data are processed while the user is still in communication with the computer. After processing the results are transmitted back to the user's terminal.

Time Sharing   This facility involves a method of conversational interactive computing and enables the compilations and program execution on a statement-by-statement basis. The user is able to solve complicated problems with rapid responses in real time by *multiprogramming*. It enables a number of remote users to gain access to a centrally located computer which provides computer power on a wider basis. The provision of time sharing facilities makes the computer available for personal use on site.

Data Telecommunications

The Post Office has established certain facilities known as *Datel Services* for the transmission of data for computing purposes.[10,11] Datel is a contraction of Data and Telecommunication services, which is a group of communication services over the telephone or telegraph lines. These services are provided in an on-line or off-line mode within the United Kingdom (Inland Datel Services) or between subscribers abroad (International Datel Services). The types of service provided are summarised in appendix B and are dependent on the mode and speed of transmission.

COMPUTER PERSONNEL

Computer processing may be carried out by gaining a direct access to the computer's facilities. This may be achieved by purchasing, renting or leasing a machine to provide in-house facilities. An in-house use of a computer means that the machine is installed on the premises of an organisation and is under their direct control.

Owning a computer involves the running of a computer department with the employment of staff performing individual specialist duties. Such a department consists of an organisation with a multiplicity of job descriptions. Although the number of persons employed by a computer department will vary from organisation to organisation, the duties to be performed will remain the same and the functions associated with a number of jobs may be carried out by one person. These specialist duties may be considered under the following headings.[12]

The services of the department will fall into three categories of work (a) computer development, (b) computer processing operations and (c) data preparation, involving the following personnel.

(a) Computer Development

> Data-processing Manager
> Systems Designers/Analysts
> Programmers

(b) Computer Processing Operations

> Operations Manager
> Chief Operator/Shift Leader
> Computer Operators
> File Librarian

(c) Data Preparation

> Data Control Section Leader
> Punch Operators

Data-processing Manager   The responsibilities of a data-processing manager may vary within different organisations. However, he may be considered to be responsible for both the computer systems development and the operation of the computer department. He is responsible to top management and his duties usually involve management and liaison. The former includes the general supervision and coordination of the activities of the department and the latter involves a responsibility to senior management for the control and monitoring of the department's progress. More specifically he would be concerned with such matters as the budgetary control of the computer department, rendering advice on the choice of equipment, organising the deployment of computer personnel within the department according to the needs of the organisation, arranging the recruitment and training of computer staff in conjunction with the personnel department and coordinating and controlling the operating functions of the department to integrate the services provided to other departments. The data-processing manager must also possess sufficient knowledge of current technology to be able to advise senior management and control staff. Some of his tasks may be carried out by consultants or, in large organisations, the tasks of high-level analysis and devlopment may be delegated to a managment information analyst who acts as a deputy.

Systems Designers/Analysts   Their responsibility is to determine the information requirements of the organisation and to prepare designs for computer-based systems to satisfy their needs. The objective is to analyse the business activity to determine what needs to be accomplished and how to achieve this end. Systems analysts will normally have a degree or professional qualification backed by some business experience.

148

Programmers   Programmers are responsible for converting the program specification into a set of instructions for the computer. They must be conversant with coding techniques and have a good knowledge of programming languages. Although the development and use of compiler languages has relegated the details of coding the importance of the programmers' function still remains. Programmers' duties have become more creative due to the greatly expanding application of computers in all areas of science and business and the need for sophisticated programs.

Operations Manager   The operations manager is responsible to the data-processing manager for the planning and day-to-day running of the computer. Computer operations must be scheduled, timed and controlled to reduce idle time and down time to a minimum. The operations manager is concerned with the efficient running of the computer and performs what could be described as a housekeeping chore.

Chief Operator/Shift Leader   Computers are required to be used as much as possible to maintain an economic benefit. This may mean that the computer must operate on a shift basis and when a multishift system is operating chief operators are appointed to deputise for the operations manager. Their duties involve the scheduling of the operations and the supervision of the working of the computer and its ancillary equipment. They are responsible for assigning duties and ensuring that records are kept that show job progress and the use of equipment. They must also ensure a good operating efficiency.

Computer Operators   These members of the staff are responsible for the actual machine operations. They assemble the input and output material and set up the machines and monitor and record the machine operations. Their duties also include the diagnosis of any malfunction of machines, which should be brought to the attention of the chief operator.

File Librarian   The file librarian is initially responsible for the security and safe keeping of computer files. This involves the classification and arrangement of the files in accordance with their contents and use. The librarian is also responsible for the inspection of files and protection against wear and tear.

Data Control Section Leader   The data control section leader is responsible for the scheduling and preparation of the workload for the computer. This involves the organisation and supervision of operating staff and the maintenance of records relating to the progress of jobs through the punch room and the maintenance of a good standard output relating to speed and accuracy.

Customer Liaison Officer   The customer liaison officer is responsible for liaison between computer user departments and the computer section. He is responsible for giving advice on computer working and answers any queries in regard to computer processing.

149

National Computing Centre Ltd

The National Computing Centre Ltd is an impartial authority in the computing world that serves industry, the professions and local and national government. It is financed by a government grant and subscriptions from its members, is independent and non-profit-making and is dedicated to promoting the use of computers. The services provided include information, advice, coordination, exchange of experience and publications.

REFERENCES

1. ——— 'Consultancy Services', *The Computer Users' Year Book*, ed. P. Grant (Computer Users' Yearbook, Brighton, 1974).
2. ——— General Brochure, Construction Control Systems Ltd.
3. ——— *Computer System General Introduction*, Local Authorities Management Services and Computer Committee.
4. ——— *Autobill: Computer Techniques Billing Systems* (Centre-File [Northern] Ltd, R.I.C.S., London, 1973).
5. J. R. Clarke, *Cosybill: Computer Techniques Billing Systems* (R.I.C.S., London, 1973).
6. ——— *The DG System – Computer Techniques Billing Systems* (Development Group of Chartered Quantity Surveyors, R.I.C.S., London, 1973).
7. P. G. Down and R. H. Sharman, *Facet – An On line Cost Control and Bill Production System: Computer Techniques Billing Systems* (R.I.C.S., London, 1973).
8. ——— *ICL PERT Users Guide 1900 Series* (Technical Publications Service, ICL, London, 1969).
9. ——— 'Service Bureaux', *The Computer Users' Yearbook*, ed. P. Grant (Computer Users' Yearbook, Brighton, 1974).
10. ——— 'Data Transmission', *The Computer Users' Yearbook*, ed. P. Grant (Computer Users' Yearbook, Brighton, 1974).
11. R. G. Anderson, *Data Processing and Management Information Systems* (MacDonald & Evans, London, 1974).
12. ——— 'Computer Personnel', *The Computer Users' Yearbook*, ed. P. Grant (Computer Users' Yearbook, Brighton, 1974).

12 ECONOMICS OF COMPUTERISATION

Computerisation has far-reaching effects on an organisation. Not only does the use of a computer create financial problems but it may also involve an organisational restructuring that results in the whole administrative machinery revolving around the computer system. There are many and varied reasons why people choose to use a computer, some of which may appear to be rather trivial. Smart[1] considers that all too often the designers and users of new techniques will be so overcome with the novelty of the approach that they will ignore the economic and/or practical aspects of the scheme. The basic considerations of cost, the effectiveness of the system and its practicability should be studied before embarking on such schemes.

When considering using computers, a feasibility study should be carried out to determine the suitability of applying computer techniques to quantity surveying problems. Each problem situation should ideally be evaluated on its own merits. The choice of the correct computer service to satisfy the needs of any user can only be answered by those having all the facts pertinent to the individual case. Choosing a computer service is like choosing a car and may often be an emotional choice at the expense of reason. To use a computer involves capital expenditure, which should be viewed in the same light as a capital investment. An adequate return must be ensured on the investment in order to make computerisation an economically viable proposition. Financial considerations may become very complicated and it is considered inappropriate to pursue such issues here. This chapter is therefore confined to certain factors and the generally accepted conclusions relating to computer usage. These involve (a) motivation, (b) computer selection and (c) the effects of computerisation.

MOTIVATION

Some of the reasons behind the motivation for using a computer may be summarised as follows.[2]

(i) To increase performance and profitability; the general use of resources, processing methods and control techniques may not be producing the best possible performance and profit. The apparent weakness in data coordination may be overcome by supplementing manual effort with automation in the form of a computer.

(ii) To resolve a problem situation; situations do arise that cannot be resolved by using existing resources and methods, due to such issues as complexity of work, shortage of trained staff or the increase in the volume of work. In this situation all possible problem solutions must be investigated as well as the potential wider use of the computer.

151

(iii) To provide a base for reorganisation; this usually occurs when the organisational structure of a business has grown with a firm's development and in consequence administration has become too complex and possibly top-heavy. Not only will the introduction of a computer be a valid reason for a complete reorganisation but it will also have the effect of streamlining and simplifying the processing methods of the organisation.

(iv) To increase prestige; this might be a possible motivation and a valid reason in the eyes of the directors or partners. In this instance the use of a computer may not prove to be a viable economic proposition.

In certain cases the use of a computer may be motivated by necessity. This might arise in a situation where the quantity surveyor is part of a large organisation that is committed to computer usage by the established policy of the management. The quantity surveyor would then appear to have little option but to help justify the use of the machine. The corporate use of the computer will help to mitigate the costs.

The progress of computer applications in the construction industry is slow and this may be attributed to a number of reasons. Organisations within the industry are comparatively small with little time, money or enthusiasm for making experimental use of new machines or new methods of application. There may also be a lack of incentive to provide a better service than others at an extra cost, particularly if these costs are difficult to recover.

The savings made through the use of computers are difficult to assess objectively; cost benefits tend to be indirect. For instance, the facilities provided by a computer system for the production of a bill of quantities can produce benefits in regard to extra information that may be made available to others. The people who benefit most from computerisation are often not the ones who bear the cost of producing the information. For example, some architects may use computers to produce extensive schedules of components that may tend to benefit the quantity surveyor and contractor more than the architect. At the same time quantity surveyors may produce various bill sortations that benefit the contractor. In this situation there seems to be little motivation; computer development and its use may be considered to be a result of necessity or of a pioneering spirit.

COMPUTER SELECTION

When introducing computer techniques certain steps must be taken to ensure that the correct method of using the computer is chosen. The problem of choosing a method of obtaining the use of a computer needs careful consideration and the best solution can only be achieved with the help of expert advice. The primary concern of any organisation should be to understand its requirements before making any definite approaches to manufacturers or consultants. Firm objectives must first be established that include the motives for introducing computer techniques and the uses to which these techniques are to be put. In establishing these objectives an investigation must be conducted into any alternative methods of achieving the same ends. A sound decision can only be made if consideration is given to the need for a computer with specific reference to the processing requirements. Consideration should also be given to the range of facilities that will best suit the processing requirements and to the organisation most likely to provide these facilities.

To specify the right computer service to satisfy the processing needs involves a feasibility study. Care must be taken to ensure that this does not form a means of justifying a preferred course of action and an objective analysis must be made to establish the real requirements. In establishing these requirements an investigation must be conducted into alternative methods of achieving the same ends. The use of the computer may be only one of a number of ways in which increased efficiency may be achieved and a cheaper method may be found. When considering using a computer a comparison should be made with the best non-computer system rather than with the system currently in use. By making these comparisons hidden deficiencies may become apparent that make the use of the computer seem more attractive. If the requirements are well defined the field of choice will be narrowed.

In order to assess the use of the computer properly a comparison must be made of the various computer services and systems that are available. Firstly, an examination should be made of the various types of computer facilities to determine their compatibility with the established requirements. Secondly, a choice should be made to provide the favoured facilities. This choice involves the selection of a suitable method of using the computer and its adherent processing techniques. These should also be considered with a view to economy.

As much information as possible must be obtained on the problems of the change-over from the manual operations currently in use to those necessary for the implementation of computer techniques. The consequences of any change regarding the organisational structure of the firm and any consequential effect on the staff must also be considered. If changes are to be implemented a timetable should be drawn up to take account of any retraining of staff. The change-over should be conducted in stages and reviewed when each stage is complete. The time scale should be sufficiently flexible to take account of any unforeseen circumstances.

No attempt will be made here to assess the specific costs of using a computer since these and individual circumstances vary. However, it is possible to study the general factors that influence the choice of a computer service to suit the needs of a prospective user such as a quantity surveyor.

Should computerisation be found to be desirable the problem initially facing the quantity surveyor will be the method of computer processing to use. Solutions may be provided by (i) acquiring a computer for in-house use, or (ii) enlisting the services of computer bureaux. These are outlined in chapter 11.

Acquisition for In-house Use

Acquiring a computer for in-house use may be achieved by purchasing a computer outright, by renting or by leasing. The question whether to buy, rent or lease is largely a matter of economics and due consideration must be given to the technological development of computers and the quantity of work passing through the office. The types and sizes of the projects to be processed should also be considered. Many organisations are reluctant to purchase a machine outright because of shortage of cash, technological changes and the fear of obsolescence.

Computer obsolescence may occur for technical, physical or economic reasons. *Technical obsolescence* is a factor that need not be considered. The computer may become obsolete due to the technological developments of the machines and the needs of the user. In this instance the user may require a machine that is technically up-to-date and in a class currently offered by manufacturers as possessing the latest innovations. Many third-generation machines are available for this reason. *Physical obsolescence* may result from a need to maintain the computer at peak performance. Usually preventive maintenance as carried out by manufacturers will keep a machine in peak condition for at least eight years. It may prove uneconomic in some situations for a user to keep a computer in good working order and yet the machine may still be of use. *Economic obsolescence* is considered to exist whenever a system is no longer cost effective in meeting its objectives. There are no definite guidelines since economic obsolescence depends on the considerations of each user and each application.

The computer falls into a price bracket that reflects the heavy cost of research and development behind the production of machines. This might appear to put computers out of the running altogether as a really economic aid to quantity surveyors. The initial cost of a machine is £50 000 upwards, with high annual maintenance costs running into four figures. The cost of computers must therefore be viewed in relation to the work they perform, in the same way that a heavy outlay is justified on contractor's plant in the right circumstances. When a machine is fully occupied the economic level is soon reached and it is at this stage that the quantity surveyor may find the in-house use of a computer to be a viable proposition. The economic factor then takes on a very significant appearance and, once programming costs have been recouped, the real cost of computer services is reduced to little more than the cost of running the machine. The private practitioner is not likely to be in the happy position of being able to have a computer standing idle for his benefit. The alternative might be to share computer staff and equipment between several offices in the locality. Even the volume of work of large practices would probably still make the exclusive use of a computer uneconomical. It may be expedient to set up a consortium for the corporate employment of computer personnel and exclusive use of a computer. Each member would then buy computer time as required. The main problem would be to organise the work in such a way as to keep staff and equipment fully occupied. A big drawback that possibly outweighs any financial advantage is the dependence on one or two key members of the computer staff. Difficulties may be experienced in the replacement of their skills at short notice in the event of illness. In such an emergency it would not be possible to call on the services of a bureau unless the data-processing techniques were exactly in accord with the requirements of the quantity surveyor. How large a firm must be to make the purchase of a computer lead to eventual savings in costs as compared with other methods of processing, is a question demanding more study and can only be answered by conducting individual feasibility studies.

The solution as to whether the computer should be purchased, rented or leased may be found in the plans for computer development. If it is possible to formulate a long-term plan and take a decision that will be valid for at least six to ten years, purchasing may be the answer. On a shorter-term basis, if changes are envisaged every four to five years, leasing may be considered the best answer. However, if more frequent changes are needed the computer should be rented. Other factors may be considered as follows.

Purchasing a computer may be considered the best possible method if there are sufficient cash resources available and if the machine is well suited to the data-processing needs of the organisation for a number of years. If the data-processing needs are for only four to five years it is considered unwise to purchase. Consideration should also be given to the surrender value of the machine in the event that disposal becomes necessary due to technological obsolescence.

Renting a computer may be considered to be the most popular method of acquiring a machine. This method may be used as an expediency on a short-term basis, particularly if the user has not been able to define his data-processing requirements fully, or if his ultimate aim is to change to another supplier. Most manufacturers provide rental agreements over two, five or seven years; the standard period appears to be two years. Renting a computer does not require an initial capital outlay and thus relieves any strain on the company's finances. The effect of any technological obsolescence is minimised and computers can be exchanged for more up-to-date models under new agreements. However, a computer can be enhanced by the addition of core storage and more powerful peripherals and in this way it may not be necessary to exchange the computer completely.

Leasing a computer may prove to be more cost effective than renting or may be a more attractive proposition than outright purchase. A number of merchant banks and finance houses offer leasing facilities that provide a flexible approach in a rapidly changing field of computer technology. They allow the user to specify the equipment he requires, then they purchase it and subsequently rent it to the user. These companies take a more active interest in the hardware and assume that the lease will have a long life. It is also anticipated that the lease rates can be maintained for several years and even after the introduction of new equipment the computer will have an important residual value. The advantage of leasing means that the user can deal directly with the manufacturer while permitting the user's accounts to show no liability against profits.

The advantages of owning a computer may be summarised as follows.[3]

(1) Tax allowances on the capital expenditure may be offset against the cost of purchasing a machine.
(2) On a long-term basis it is usually cheaper to buy than to rent or lease.
(3) Disposal of the computer at any time is made easier if it is owned. Rental and leasing agreements are for fixed periods.
(4) The disposal of the computer realises a cash inflow that creates a fund that may be offset against purchase and is an asset on the balance sheet.
(5) A charge is usually made for excess running hours if the computer is rented. This charge will not be incurred if the computer is owned.
(6) There are no fixed monthly cash outflows of the kind associated with rental or leasing agreements.

The disadvantages of owning a computer may be summarised as follows.

(1) Technological obsolescence may result in having to dispose of the computer prematurely.
(2) Initial cash outflow may create a liquidity problem for a company and a liability incurred in regard to interest on loans.
(3) If the computer is disposed of prematurely there will be capital losses incurred.
(4) Funds may be restricted for alternative investments.
(5) The optimum benefits may not be secured if the computer is being used without operating experience.

The acquisition of a computer will involve certain costs consisting of initial costs of implementing a computer system and annual operating costs. The initial costs will involve the construction of new premises or the conversion of existing premises with air-conditioning equipment to accommodate the computer, the provision of hardware and software, staff training and the costs of converting master files from the existing system to the computer system. Annual operating costs will include general expenses involving the purchase of stationery, input/output media and the like. Staff salaries, general administrative expenses and other related costs will also be incurred.

Quantity surveyors in the public service are rapidly acquiring the in-house use of computers of various sizes. In the course of time there may be few such surveyors who will not have access to a computer within their own organisations. Government departments, Local Authorities and large commercial organisations are in a particularly favourable position. Whether to buy, rent or use a bureau may hinge on the volume of work and the charges made for the services rendered.

Organisations most able to take advantage of in-house computer services are large organisations using the computer for many purposes. The only method open to small organisations wishing to use the computer is to enlist the services of external agencies. This may be achieved by using a computer bureau or obtaining computer power by the use of a remote data terminal.

Computer Bureaux Services

Certain factors may influence the choice of external computer services in preference to in-house facilities. For instance, the amount of computer experience available may be a deciding factor which, if limited, may prompt a decision in favour of a bureau in order to gain experience without any high capital investment. It may be considered better to gain initial experience in the effects of computerisation without the burden of having to control the operations. Another factor to be considered is the amount of work passing through the office. This should be sufficient in quantity to warrant the acquisition of a computer, otherwise the use of bureaux services may be the only feasible alternative. Problems may arise involving peak and slack periods that would result in a computer standing idle. In this situation bureaux services or data terminals should be considered. If the apparent difference in the costs of using a bureau and owning a computer is only marginal it may be considered better to use a bureau in order to avoid the problems of owning a computer.

The use of a computer bureau may depend on the bureau's ability to provide an application package that is well suited to the processing needs of the quantity surveyor. Application packages usually lend themselves to a generalised solution. The first impression is that a package provides a natural and cheap solution. However, the economic benefits seem to disappear when the computer user compares the working of the package with his detailed requirements. There is a tendency to produce packages that will answer the well-established problems associated with the system but will lose sight of the individual needs and objectives of the user. A general approach often results in many packages being written in such a manner as to preclude easy modification by a potential user. Notwithstanding this criticism, application packages are playing an important role in providing computer facilities for quantity surveying services. Many packages have been developed in close liaison with quantity surveyors (see chapter 11). With the installation of more powerful computers by computer bureaux, progress has been made in the field of direct access data transmission services with the location of input and output peripherals in the quantity surveyor's office. In this way the quantity surveyor may hire computer time and gain direct access to the computer. The charges for hired time are usually on an hourly basis. The costs of using external services are more tangible and easily assessed and may merely involve the payment of a fee for processing.

EFFECTS OF COMPUTERISATION

The economic benefit of the computer to the quantity surveyor may be appreciated by examining the effects that computerisation has had on the profession. The computer's ability to sort information is probably more significant in data-processing for the construction industry than the speed at which it performs calculations. Once information has been converted into a form that can be accepted by the machine it makes good economic sense for it to perform all the necessary calculations needed at the same time as sorting information. Merely to use the computer to calculate, sort and generate information in the form of bills of quantities is not to take full advantage of the machine. Certain orthodox bills can be produced economically and quickly by manual methods thereby justifying this method of preparation. However, additional information may be produced by a computer, which thus offers a bonus. The use of a computer enables the reproduction of information in various bill formats such as 'elemental', 'locational' and 'operational' bill sortations. The more elaborate systems exploit this ability and more sophisticated bills of quantities can be produced which help with other tasks. The uses to which a bill of quantities can be put affect the efficiency with which the quantity surveyor fulfils his several roles.

The manual process of taking-off may be reduced slightly as a result of techniques that can be adopted by using a computer system that provides certain facilities such as short coding, unit quantities and automatic coding techniques.[4,5] Any reduction in the time taken to perform necessary manual operations will provide financial benefits. However, it must be appreciated that a great deal of time is spent on the creation of short code and unit quantity libraries and savings will only accrue over a period of time.

Data produced by the quantity surveyor may be made available for such processes as estimating, cost planning and control, project planning, construction planning, ordering, financing and valuations. However, there is an economic problem of processing the pieces in order to supply the construction team as a whole and the contractor with the full scope of data prepared by the quantity surveyor at the planning stage. The processing of data originated by the taker-off is a potential source of error if carried out manually. The use of the computer is aimed at speeding up operations and reducing errors to a minimum by eliminating any unnecessary human operations.

The basic principles, suggested by a working party of the Quantity Surveyors' Committee, that should be considered and accepted before any change is recommended in the traditional methods of preparing bills of quantities, are as follows.[6]

(a) The standard of service to the client or industry should not be reduced by virtue of the use of mechanical aids.
(b) The adoption of any mechanical aid must not involve any extra expense that would lead to an increase in professional fees at the time.
(c) The mechanical aid must be accurate and reliable and errors must be human- and not machine-generated.
(d) The surveyor must be able to check the accuracy of any operation carried out by the mechanical aid.
(e) The use of any mechanical aid should show a saving in time required for the production of bills of quantities and final accounts.
(f) The taker-off should be disturbed as little as possible and any procedure that would lessen his efficiency should be avoided.
(g) Rewriting original dimensions and descriptions and the transfer of figures should be reduced to a minimum.
(h) Any necessary training in the use of mechanical aids must not jeopardise the efficient training of future surveyors.
(i) The working party, in accepting electronic computers and data processing equipment as reliable and accurate mechanical aids, consider that the surveyor should have control of the information fed into the machine either by carrying out his own coding or by checking the coding carried out by others.
(j) If the surveyor employs an outside mechanical aid service he must remain responsible for and satisfy himself as to the accuracy of the resultant product.

The saving of time by the use of a computer does not necessarily mean the displacement of manpower in the quantity surveyor's office but rather supports the quantity surveyor's changing role with more opportunities for advancement. It would not be true to say that the quantity surveyor no longer prepares bills of quantities by virtue of the innovation of computer processing techniques. What computerisation really means is that the old and well-tried methods are being replaced by more modern sophisticated techniques. A considerable amount of drudgery is being taken out of the quantity surveyor's work and the expendable time saved by the removal of such manual procedures as abstracting and billing. The time saved may be usefully employed on the major functions of taking-off, measuring and agreeing final accounts and the development of cost consultancy services. This is at a time when staff are hard to get and the quantity surveyor is becoming more professional in his capacity as financial adviser in the design team. The changing role of the quantity surveyor is not necessarily due to these new techniques, but at least computers do play their part.

Assessing the true benefit of computers in quantity surveying is very difficult and requires a more extensive study than this book is able to supply. There are many and varied opinions as to the benefit of computerisation, some for and some against. The author can only leave it to the reader who, it is hoped, is in a more enlightened position and better able to judge in the light of his own experience. The choice of computer processing is like selecting a mode of transport. It depends on the journey, its length and your own preferences. Whether you walk, use public transport or buy a car depends on your ability to drive, the length of the journey, and how often the journey is made.

REFERENCES

1. D. A. Smart, *Bureau Services Computer Techniques Billing Systems* (R.I.C.S., London, 1973).
2. T. F. Fry, *Computer Appreciation* (Butterworth, London, 1970).
3. R. G. Anderson, *Data Processing and Management Information Systems* (MacDonald & Evans, London, 1974).
4. —— *Computer System Handbook*, Parts 1M and 3M, Consortium of Local Authorities Special Programme: Central Development Group.
5. —— *The DG System – Computer Techniques Billing Systems* (Development Group of Chartered Quantity Surveyors, R.I.C.S., London, 1973).
6. Royal Institution of Chartered Surveyors: Quantity Surveyors Committee, 'Mechanical and Other Aids to Quantity Surveyors', *Chart. Surv.*, 95 (1962) pp. 168–72.

APPENDIX A UNIT QUANTITY LIBRARY

Item	Quantity	Description	Master Code	Parameters
1.	1.16 M	Precast concrete; normal; mix 1:2:4 20 mm aggregate; lintels; bedding in cement mortar (1:3); 102 x 75; reinforced one 10 mm bar	542 L	S − −−
2.	2.32 M	Precast concrete; normal; mix 1:2:4 20 mm aggregate; lintels; bedding in cement mortar (1:3); 102 x 75; reinforced one 10 mm bar	542 L	D − −−
3.	9.60 M2	Selected common bricks; in cement mortar (1:3); half brick thick; flush pointing one side; walls; stretcher bond	542 L	S O CB
4.	5.47 M2	Selected common bricks; in cement mortar (1:3); half brick thick; flush pointing one side; walls; stretcher bond	542 L	S C CB
5.	11.21 M2	Selected common bricks; in cement mortar (1:3); half brick thick; flush pointing one side; walls; stretcher bond	542 L	D O CB
6.	7.09 M2	Selected common bricks; in cement mortar (1:3); half brick thick; flush pointing one side; walls; stretcher bond	542 L	D C CB
7.	3.85 M2	Selected common bricks; in cement mortar (1:3); half brick thick; flush pointing both sides; walls; stretcher bond	542 L	D − CB
8.	13.45 M2	Solid concrete blocks; 100 mm thick; walls or partitions	542 L	S O BL
9.	9.32 M2	Solid concrete blocks; 100 mm thick; walls or partitions	542 L	S C BL

Item	Quantity	Description	Master Code	Parameters
10.	15.06 M2	Solid concrete blocks; 100 mm thick; walls or partitions	542 L	D O BL
11.	10.94 M2	Solid concrete blocks; 100 mm thick; walls or partitions	542 L	D C BL
12.	5.50 M	Bonding ends to brickwork; 100 mm blockwork	542 L	S O BL
13.	3.32 M	Bonding ends to brickwork; 100 mm blockwork	542 L	S C BL
14.	8.25 M	Bonding ends to brickwork; 100 mm blockwork	542 L	D O BL
15.	6.07 M	Bonding ends to brickwork; 100 mm blockwork	542 L	D C BL
16.	4.25 M	100 mm Blockwork against soffits	542 L	S O BL
17.	2.75 M	100 mm Blockwork against soffits	542 L	S C BL
18.	7.00 M	100 mm Blockwork against soffits	542 L	D O BL
19.	5.50 M	100 mm Blockwork against soffits	542 L	D C BL
20.	5.17 M	Bedding in cement mortar (1:3); wood frames or sills	542 L	S – BL
21.	10.54 M	Bedding in cement mortar (1:3); wood frames or sills	542 L	D – BL
22.	5.17 M	Bedding in cement mortar (1:3); wood frames or sills; pointing one side	542 L	S – CB
23.	10.54 M	Bedding in cement mortar (1:3); wood frames or sills; pointing one side	542 L	D – CB
24.	1 NO	Flush door; skeleton core; 44 x 826 x 2040 mm; plywood facing both sides; hardwood lipping long edges	542 L	S – ––
25.	2 NO	Flush door; skeleton core; 44 x 826 x 2040 mm; plywood facing both sides; hardwood lipping long edges	542 L	D – ––
26.	5.32 M	Softwood; door frame; 100 x 75 mm rebated and moulded labours	542 L	S – ––

Item	Quantity	Description	Master Code	Parameter
27.	10.64 M	Softwood; door frame; 100 x 75 mm rebated and moulded labours	542 L	D − −−
28.	1 Pr	To softwood; mild steel; butts; 75 mm	542 L	S − −−
29.	2 Pr	To softwood; mild steel; butts; 75 mm	542 L	D − −−
30.	1 NO	To softwood; rim locks; ref RL123	542 L	S − −−
31.	2 NO	To softwood; rim locks; ref RL123	542 L	D − −−
32.	6 NO	To softwood; wrought iron; galvanised cramps; 3 x 25 x 250 mm girth; bent once; holed twice	542 L	S − −−
33.	12 NO	To softwood; wrought iron; galvanised cramps; 3 x 25 x 250 mm girth; bent once; holed twice	542 L	D − −−
34.	12.46 M2	Mortar; cement and sand (1:3); steel trowelled; to walls; over 300 mm wide	542 L	S − BL
35.	24.81 M2	Mortar; cement and sand (1:3); steel trowelled; to walls; over 300 mm wide	542 L	D − BL
36.	10.24 M2	Plaster; first coat gypsum plaster undercoat and sand; finishing coat gypsum plaster finishing steel trowelled; to walls over 300 mm wide	542 L	S O −−
37.	5.81 M2	Plaster; first coat gypsum plaster undercoat and sand; finishing coat gypsum plaster finishing steel trowelled; to walls over 300 mm wide	542 L	S C −−
38.	13.95 M2	Plaster; first coat gypsum plaster undercoat and sand; finishing coat gypsum plaster finishing steel trowelled; to walls over 300 mm wide	542 L	D O −−
39.	9.55 M2	Plaster; first coat gypsum plaster undercoat and sand; finishing coat gypsum plaster finishing steel trowelled; to walls over 300 mm wide	542 L	D C −−
40.	3.62 M2	Painting one coat lead free wood primer; two undercoats, one coat alkyd based paint full gloss finish; wood surfaces; general surfaces; over 300 mm girth	542 L	S − −−

162

Item	Quantity	Description	Master Code	Parameter
41.	7.32 M2	Painting one coat lead free wood primer; two undercoats, one coat alkyd based paint full gloss finish; wood surfaces; general surfaces; over 300 mm girth	542 L	D – ––
42.	5.17 M	Painting one coat lead free wood primer; two undercoats, one coat alkyd based paint full gloss finish; frames or the like; over 100 mm not exceeding 200 mm girth	542 L	S – ––
43.	10.34 M	Painting one coat lead free wood primer; two undercoats, one coat alkyd based paint full gloss finish; frames or the like; over 100 mm not exceeding 200 mm girth	542 L	D – ––
44.	10.24 M2	Painting three coats emulsion wall finish to plaster	542 L	S O – –
45.	5.81 M2	Painting three coats emulsion wall finish to plaster	542 L	S C – –
46.	13.95 M2	Painting three coats emulsion wall finish to plaster	542 L	D O ––
47.	9.55 M2	Painting three coats emulsion wall finish to plaster	542 L	D C ––

APPENDIX B INLAND DATEL
SERVICES

Datel 100 is a service providing data communication facilities over a private telegraph circuit or public telex network. The operation of private telegraph circuits provides a means of data transmission on a point-to-point basis for the exclusive use of the customer. Transmission speeds of 50 to 110 bits (binary digits) per second can be assured depending on the circuits. The operation of the public telex network is by means of teleprinters at transmission speeds of 50 bits per second. The terminal equipment that may be used in association with this system may be (a) teleprinters fitted with a tape punching attachment for the preparation and reception of data in a punched tape medium, (b) a tape reperforator for receiving data as fully punched perforations on paper tape, which is not available in telex communications, (c) an automatic transmitter that is a reading device for punched paper tape, (d) an error-detection unit, (e) a switching device to enable privately owned equipment to be connected to telex or telegraph private circuits for transmitting data.

Datel 200 is a service providing serial data transmission over public telephone networks or private speech circuits. Speeds of up to 200 bits per second can be assured with a possible 300 bits per second as a maximum. The system is operated over public telephone networks by setting up a normal telephone call and switching over to a modem and terminal devices when contact has been established. Incoming and outgoing transmissions may be made simultaneously if required. These facilities can also be provided on internal circuits that are used exclusively for speech or data transmission within different departments of an organisation. The data transmitters may be switched to operate on line or off line by means of a special switching device.

Datel 400 is a service that is concerned with telemetry applications and facilitates the transmission of digital or analog data over public telephone networks or private circuits. The service may be operated over telephone lines with or without a telephone attachment. If no telephone is fitted all calls will be answered automatically. However, if a telephone is fitted all calls may be answered manually or automatically by means of a locking button. This means that remote unmanned stations can be 'polled' over the public telephone network with the subsequent transmission of data to the central control station. Transmission speeds can be assured of up to 600 bits per second for digital data and 300 cycles per second for analog data.

Datel 600 facilitates the serial transmission of digital data within two speed ranges of 600 bits per second over public telephone networks and 1200 bits per second over private circuits. A supervisory channel may also be provided as an option with a transmission speed of 75 bits per second. A special switching device can be supplied enabling the data transmitters to be used in an on-line or off-line situation to meet the needs of the user.

Datel 2400 is a facility providing a synchronous both-way transmission of serial binary data at a rate of 2400 bits per second over private circuits and at an assured rate of 600 bits per second over the public telephone network. Unattended answering facilities may be provided when data terminals are connected to the public switched network for alternative working similar to the facilities provided by the Datel 400 service.

Datel 2400 Dial-up service provides facilities over the inland public switched telephone network only, at a transmission rate of 2400 bits per second. This facility provides a synchronous communication and is limited to data transfer in either direction alternately. If the user so desires he may switch to an alternate mode which is identical to the fall back mode of the Datel 2400 service at transmission rates of 600 and 1200 bits per second.

Datel 48K provides facilities for a transmission of 40.8K, 48K or 50K bits per second over a wideband circuit. This is a circuit that provides bulk communication facilities at a very high speed. The use of this type of circuit means that the bandwidth can be split into different circuits for speech or data transmission.

Dataplex Services make provision for a number of data calls transmitted simultaneously to the same computer centre over a single circuit. This means that several remote terminals in the same locality (for example, telephone area) can gain access to the computer at the same time. The facilities offered by the Post Office involve a system of frequency division multiplexing, which provides a number of channels for communication. *Dataplex System 1* was designed to cater for the Datel 200 users and provides speeds up to 110 bits per second with a choice of 6 or 12 channel working. *Dataplex System 2* has a varying transmission speed up to 1200 bits per second and uses a time division multiplexing that is divided into a number of time slots to provide a number of low-speed channels capable of a relatively high information transfer rate.

International Datel Services

The international Datel services have been developed for use in a number of foreign countries and cover the Datel 100, Datel 200 and Datel 600 services over the telex systems or public telephones.

International Datel 100 service is used over the public telex network at transmission speeds of 50 bits per second. The operation of the service may be set up and cleared according to the normal telex procedures.

International Datel 200 is designed to provide a service for the transmission of digital data in both directions simultaneously. The speed of transmission is up to 200 bits per second over international telephone circuits.

International Datel 600 provides for the transmission of serial digital data within a speed range of 600 to 1200 bits per second over the public telephone network.

APPENDIX C GLOSSARY OF TERMS

Abacus An early digital counting device using sliding beads or counters on rods

Access time The time taken to retrieve data from memory for processing

Accumulator A section in storage to which data are transferred for arithmetical manipulation

Activity A concept of production planning, regarding the application of resources to operations

Address A unique reference to a character location in memory

ALGOL An abbreviation of ALGOrithmic Language or ALGebraic Orientated Language that is a computer processing language for presenting numerical procedures

Algorithm A step-by-step procedure for achieving a solution

Alphameric A contraction of alphanumeric

Alphanumeric Characters consisting of numbers and letters of the alphabet

Analog computer A device that measures the quantity of some variable physical condition; a measure of 'how much'

Arithmetic unit The part of the central processor that performs the arithmetic tasks associated with digital processing

Assembler A computer program that translates a symbolic processing language program into machine instructions item for item

Backing store Computer storage held by peripheral devices to supplement main frame core storage

Base The total number of different digits that form a numbering system

BASIC A computer programming language primarily designed for interactive (conversational) computing

Basic cycle The time taken to complete a set of operations for the execution of an instruction

Batch An assemblage of related data

Batch processing The processing of information in groups of related data

Binary A numbering system using the base 2

Binary code An arrangement of symbols employing binary digits with a base of 2 (ones and zeros)

Bit An abbreviation for binary digit; the smallest unit of information in the computer

Block A set of computer words on magnetic tape arranged in a sequential order and handled as a unit

Bus A path over which information is transferred from source to destination

Byte A group of binary digits treated by the computer for operational purposes as a unit

Category A convenient grouping of subjects included in a set of documents, including subject headings, the specification of the document and its address on a sheet in a file, or on a magnetic tape, and the like

Central Processing Unit (CPU)   The main frame component of an electronic digital computer that houses the control, arithmetic and memory units

Channel   See *Track*

Character   A symbol used to express information

COBOL   An abbreviation of Common Business Oriented Language that is a computer programming language for commercial and business applications

Code   A system of symbols used for representing information

Code key   A set of rules governing encoding and decoding

Coding   The process of translating data or information into a coded form

Compiler   A computer program that translates a symbolic processing language program into machine instructions; it is problem-orientated rather than machine-dependent and generates a number of machine instructions from one program instruction

Computer console   A device through which the operator can communicate with the computer; the device usually contains a typewriter keyboard

Control section   The part of the central processor that controls the computer's automatic processing

Core store   The main storage unit in the central processor

Cycle time   The least time required to complete an action in a computer

Cylinder of data   Information recorded in the same relative plane on each of the recording surfaces of a magnetic disc pack

Data   A general term used to denote the basic elements of information for computer processing

Data bank   A centralised computer storage facility

Data file   A collection of data words

Data records   Data pertaining to a particular subject, for example, a punched card containing information relating to an item of measured work in the take-off

Data word   An ordered set of characters forming a meaningful entity and transferred by the computer as a unit of information

Decoding   The process of transforming coded information into plain language

Design realisation   A function of the building process that lies between the choice of principle structural techniques and materials and the commencement of planning for construction work

Digit   A symbol expressing a discrete integral value in a given base to represent all the quantities that occur in a problem or a calculation

Digital   The use of discrete integral numbers in a scale of notation to represent variables relating to a problem

Digital computer   A device that handles data in a numerical or digital form; a measure of 'how many'

Digitiser   A device for converting analog measurements into a digital form

Disc storage   The storage of information on the surface of magnetic discs

Documentation   A written or graphic record of data-processing

Drum storage   A storage device of drum-like appearance on which data are stored magnetically

EDP (electronic data-processing)   The technology adopted for data processing

167

that uses electronic circuits and components to represent data by means of pulses of electricity flow

Element (*building*) A simple recognisable part of a building that corresponds to a design function

Element (*computer*) A component part of computer hardware

Encoding See *Coding*

Executive A digital computer programming routine that coordinates, directs, modifies and controls other subordinate routines. Generally associated with assemblers and compilers

Feasibility study A preliminary analysis of costs and operations to provide a basis for a decision

Feedback The return of output data to input

Field A specified area of a record containing a particular category of data; significant groups of symbols of a code

File The collection of related records that forms a unit

Flowchart A graphical representation of procedures using symbols to represent the operations

FORTRAN An abbreviation of FORmula TRANslator; a computer programming language for scientific and mathematical applications

Generation The term used to differentiate the technology of a computer

Hardware The devices or components of a computer

High-level language A computer programming language that resembles English or mathematical notations, single statements of which may be compiled into several machine instructions; it is problem-orientated rather than machine-dependent

Hybrid computer A device that possesses analog and digital characteristics and performs calculations in an analog fashion with digital storage

Information An aggregation of data

Input The act of entering data into a data-processing system

Instructions The orders or commands given to the computer to perform specific tasks

Inter-block gap A space on magnetic tape or disc separating the data files

Interpreting The function of reading the punched holes on a card and printing the relevant character on the same

Jump An instruction used to alter the normal sequence control of the computer and which specifies the location and directs the computer to the next instruction

K bits A unit measure of binary digits, which indicates the capacity of core storage, for example 1024 bits

Keyboard A device for encoding data by depressing keys, that causes the generation of a coded element

Location A position in the computer's memory that permits the storage of one computer word

Looping The repetition of a group of instructions in a routine

Low-level language A computer programming language that is machine-orientated and has a one-to-one relationship between written instructions and machine instructions

Macro code   A coding system that reduces groups of computer instructions into single code words

Main storage   The primary storage of the computer in relation to backing storage; see *Core store*

Memory   The area in the computer in which data are stored

Microsecond   One-millionth of a second

Millisecond   One-thousandth of a second

Minicomputers   Small computers

Multiprogramming   A technique of overlapping or interleaving the execution of numerous routines or programs simultaneously

Nanosecond   One-billionth of a second

Network analysis   A technique used to represent the interrelationship of activities and events by means of an arrow diagram

Notation   A manner of representing numbers

Numeric code   A system of coding using numbers

Object program   A computer program in machine-coded terms produced by assembling and compiling routines from a 'source' program

Off-line   Reference to devices not directly connected to the computer and under the control of the central processor

On-line   Reference to devices that are connected to the central processor and under its control

Output   The term used for describing the data or information leaving the computer

Peripheral   Appertaining to hardware outside the central processing unit

Program   A series of instructions given by a programmer setting out the operations to be performed

Programme   A plan of intended proceedings

Programming   The art of producing a series of instructions to direct the operation of the computer

Punched card   An input medium consisting of holes punched on to a card

Random access storage   A device that facilitates the direct access of stored data

Read   To acquire information from some form of computer storage

Read/write head   A device consisting of a magnetic coil that is capable of magnetising a tiny spot on the surface of magnetic storage media

Real time processing   A technique used to provide a rapid response computer processing

Record   Information relating to a particular item

Register   A character location in memory

Reproducer   A device used for copying data on punched cards

Routine   A set of coded instructions that directs the computer to perform some desired operation

Run   A single performance of a computer program

Serial access   A storage device that permits the access of stored data by means of a sequential search

Set up   The preparation of pieces of equipment for operation to solve a particular problem

Simultaneity   A technique involving the overlap of input and output operations with the calculating processes to form concurrent operations; the simultaneous operation of peripheral devices is automatically monitored to guarantee immediate response to their demands

Software   A term used that is complementary to hardware and refers to all the programs used by the computer for operating a computer system and computer processing

Sortation   The arrangement of bills of quantities into various formats such as trade bills, elemental bills, activity bills, operational bills and the like

Time sharing   A technique by which users gain access to the central processor on a timed rota basis

Track   The term referring to the individual rows of bits

Verifier   A device used to check the accuracy of punching

Volatile storage   A storage medium in which information is destroyed when the power is removed

Word   See *Data word*

Working storage   A portion of core store reserved for intermediate results in processing

Write   The process of transferring information to an output medium

Xerography   A dry copying process involving a photoelectric discharge of an electrostatic charge on the plate

INDEX

174